# Design and Destiny

**Basic Bioethics**
Glenn McGee and Arthur Caplan, editors

Peter A. Ubel, *Pricing Life: Why It's Time for Health Care Rationing*

Mark G. Kuczewski and Ronald Polansky, eds., *Bioethics: Ancient Themes in Contemporary Issues*

Suzanne Holland, Karen Lebacqz, and Laurie Zoloth, eds., *The Human Embryonic Stem Cell Debate: Science, Ethics, and Public Policy*

Gita Sen, Asha George, and Piroska Östlin, eds., *Engendering International Health: The Challenge of Equity*

Carolyn McLeod, *Self-Trust and Reproductive Autonomy*

Lenny Moss, *What Genes* Can't *Do*

Jonathan D. Moreno, ed., *In the Wake of Terror: Medicine and Morality in a Time of Crisis*

Glenn McGee, ed., *Pragmatic Bioethics,* 2d edition

Timothy F. Murphy, *Case Studies in Biomedical Research Ethics*

Mark A. Rothstein, ed., *Genetics and Life Insurance: Medical Underwriting and Social Policy*

Kenneth A. Richman, *Ethics and the Metaphysics of Medicine: Reflections on Health and Beneficence*

David Lazer, ed., *DNA and the Criminal Justice System: The Technology of Justice*

Harold W. Baillie and Timothy K. Casey, eds., *Is Human Nature Obsolete? Genetics, Bioengineering, and the Future of the Human Condition*

Robert H. Blank and Janna C. Merrick, eds., *End-of-Life Decision Making: A Cross-National Study*

Norman L. Cantor, *Making Medical Decisions for the Profoundly Mentally Disabled*

Margrit Shildrick and Roxanne Mykitiuk, eds., *Ethics of the Body: Post-Conventional Challenges*

Alfred I. Tauber, *Patient Autonomy and the Ethics of Responsibility*

David H. Brendel, *Healing Psychiatry: Bridging the Science/Humanism Divide*

Jonathan Baron, *Against Bioethics*

Michael L. Gross, *Bioethics and Armed Conflict: Moral Dilemmas of Medicine and War*

Karen F. Greif and Jon F. Merz, *Current Controversies in the Biological Sciences: Case Studies of Policy Challenges from New Technologies*

Deborah Blizzard, *Looking Within: A Sociocultural Examination of Fetoscopy*

Ronald Cole-Turner, ed., *Design and Destiny: Jewish and Christian Perspectives on Human Germline Modification*

# Design and Destiny

## Jewish and Christian Perspectives on Human Germline Modification

edited by Ronald Cole-Turner

The MIT Press
Cambridge, Massachusetts
London, England

For information about special quantity discounts, please email special_sales@mitpress.mit.edu.

This book was set in Sabon by SNP Best-set Typesetter Ltd., Hong Kong. Printed on recycled paper and bound in the United States of America.

Library of Congress Cataloging-in-Publication Data

Design and destiny : Jewish and Christian perspectives on human germline modification / edited by Ronald Cole-Turner.
p. ; cm. – (Basic bioethics)
Includes bibliographical references and index.
ISBN 978-0-262-03373-2 (hardcover : alk. paper) – ISBN 978-0-262-53301-0 (pbk. : alk. paper) 1. Medical genetics–Religious aspects–Judaism. 2. Medical genetics–Religious aspects–Christianity. 3. Genetic engineering–Religious aspects–Judaism. 4. Genetic engineering–Religious aspects–Christianity. I. Cole-Turner, Ronald, 1948– II. Series.
[DNLM: 1. Genetic Engineering–ethics. 2. Bioethics. 3. Christianity. 4. Germ Cells. 5. Judaism. 6. Religion and Medicine. WB 60 D457 2008]
RB155.D42 2008
201′.666065–dc22

2007032376

10 9 8 7 6 5 4 3 2 1

# Contents

Series Foreword vii
Acknowledgments ix

1 **Religion and the Question of Human Germline Modification** 1
Ronald Cole-Turner

2 **Judaism and Germline Modification** 29
Elliot N. Dorff

3 **The Roman Catholic Magisterium and Genetic Research: An Overview and Evaluation** 51
Thomas A. Shannon

4 **A Traditional Christian Reflection on Reengineering Human Nature** 73
H. Tristram Engelhardt, Jr.

5 **Germline Gene Modification and the Human Condition before God** 93
Nigel M. de S. Cameron and Amy Michelle DeBaets

6 **Human Germline Therapy: Proper Human Responsibility or Playing God?** 119
James J. Walter

7 **Germline Genetics, Human Nature, and Social Ethics** 145
Lisa Sowle Cahill

8 **Freedom, Conscience, and Virtue: Theological Perspectives on the Ethics of Inherited Genetic Modification** 167
Celia Deane-Drummond

**9 Religion, Genetics, and the Future** 201
Ronald Cole-Turner

Suggestions for Further Reading 225
Contributors 229
Index 231

# Series Foreword

We are pleased to present the twenty-third book in the series Basic Bioethics. The series presents innovative works in bioethics to a broad audience and introduces seminal scholarly manuscripts, state-of-the-art reference works, and textbooks. Such broad areas as the philosophy of medicine, advancing genetics and biotechnology, end-of-life care, health and social policy, and the empirical study of biomedical life are engaged.

Glenn McGee
Arthur Caplan

# Acknowledgments

Chapter 6 by James J. Walter, entitled "Human Germline Therapy: Proper Human Responsibility or Playing God?" is based on material previously published under the title "'Playing God' or Properly Exercising Human Responsibility? Some Theological Reflections on Human Germ-Line Therapy," in *New Theology Review: An American Catholic Journal for Ministry*, volume 10, number 4, November 1997, pp. 39–59, whose permission to reuse this material is acknowledged with appreciation.

Chapter 8, "Freedom, Conscience, and Virtue: Theological Perspectives on the Ethics of Inherited Genetic Modification" by Celia Deane-Drummond, is based on material she first presented as a lecture at Garrett Evangelical Theological Seminary on April 24, 2003, entitled "Forbidden Knowledge: A Theologian's View."

The idea of this book first arose in the late 1990s when I was a member of the working group assembled by the Program of Dialogue on Science, Ethics and Religion of the American Association for the Advancement of Science as part of a project on Human Inheritable Genetic Modifications. For the opportunities provided by this project, I am grateful to the AAAS and its staff, to other members of the working group, and to the Greenwall Foundation for its support of the project.

# 1

# Religion and the Question of Human Germline Modification

Ronald Cole-Turner

Advances in biotechnology are bringing us closer to the day when human beings will engineer specific genetic changes in their offspring. Some see this as the ultimate in human folly. They fear that parents, merely by knowing they have the option to design the child they want, will forget how to love the child they are given. Others see such genetic modification as a logical extension of medicine, consistent with basic human values and parental love.

Should we encourage the development of this technology and embrace it when it arrives? Should we human beings modify our offspring through genetic modification of the human germline? Pondering these possibilities, Hans Jonas asked: "Whether we have the right to do it, whether we are qualified for that creative role, is the most serious question that can be posed. . . . Who will be the image-makers, by what standards, and on the basis of what knowledge?"[1] With his questions, Jonas calls our attention not so much to technology as to our vision of a technologically modified humanity. What does it mean to be human, to be the sort of human that uses these technologies, or to be a human being upon whom they are used? What are the limits of human action, and who or what is guiding the process?

Like Jonas, the contributors to this book call our attention not to technology but to humanity. They draw upon the resources of traditional Judaism and Christianity to reflect on the meaning and destiny of human life, the values and principles that guide human behavior, and the meaning of our use of medicine and technology to maintain our health and to improve our condition.

A public conversation about germline modification has already begun. So far, however, the partners in the conversation are largely limited to scholars in fields such as bioethics and public policy. Aside from isolated comments and momentary worries about the dangers of "designer babies" or films such as *GATTACA*, the wider public has not been involved. What is needed is a public discussion that is broadly participatory and richly informed, building on but actively expanding the current discussion, which has largely "been confined to elite governmental commissions or scholarly groups."[2]

One way to expand the conversation and to engage the public is to approach the question of human germline modification in religious terms. Religion is the language of morality for many if not most human beings, even in late modernity. Beyond its capacity to reach a wider public, however, religion introduces something new precisely by reintroducing something old. By drawing attention to rich traditions of belief and morality, religious voices enrich the debate, adding complexity, multidimensionality, and counterintuitive thinking. For that reason, and not for mere political sensitivity, religious scholars are often invited to participate in public discussions of science, technology, and public policy. This book, too, is based on the hope that religious voices might deepen the public conversation about human germline modification, taking it to new dimensions of reflection on the meaning of our humanity.

## Is This Book Really Needed?

Even so, many may think that a book on religious perspectives on human germline modification is not needed. One reason is that the technical feasibility of human germline modification is still far off in the distant future. Overcoming the scientific and technological barriers standing its way will require decades at the very least, it is said, and once the technical possibility is clearly in sight (if ever), there will be plenty of time to debate the wisdom and morality of the use of the technology.

Another reason why some may think this book is not needed is because religion really has no legitimate or constructive role to play in public discussions about science and technology. In a secular and pluralistic age, public conversation about the future of human nature must be grounded in philosophy, not in religious doctrines. Of course, even in our secular era, religion shows no sign of dying. But the religions disagree with each other on just about everything, and they cannot possibly all be right. No single denomination or religious institution can claim to speak for more than a minority. By contrast, it is said, philosophy is universal in its assumptions and therefore deserves the sort of global respect that religion can never attain. A third objection is that no one needs a book to tell them that religious leaders and scholars are strongly opposed to human germline modification. This is common knowledge, or so it is thought.

All three objections, however, are based on misunderstandings. The truth is that a public discussion of human germline modification is timely because the technology is closer than many think, that each religion and every philosophy are all limited in the power to persuade more than minorities, and that a surprising range of religious scholars and leaders actually endorse some forms of human germline modification. Each of these points deserves a brief comment.

### The Discussion Is Timely

Germline modification of nonhuman species has been under way for more than twenty years and is becoming routine in areas such as agriculture and biomedical research using animals. Researchers have created transgenic or germline-modified sheep, mice, rats, and even a primate, the rhesus monkey.[3] The techniques that are used on nonhuman animals such as sheep or mice involve the production and destruction of many embryos. These techniques are universally regarded as ethically unacceptable for use on human beings.

Research is currently under way on a wide range of technologies that might change this situation. No one can predict exactly when or how these technical hurdles might be overcome, but researchers in the field generally believe that given enough time, the technology of germline modification will develop to the point where the techniques themselves pose no insurmountable ethical obstacle. In other words, some day human germline modification will be safe and achievable by techniques that are generally regarded as ethical for use on human beings. When that happens, the moral question of the wisdom of using the technology will be squarely before us.

While no one can predict how long it will take for research to bring us to this point, it is clear that recent research has advanced rapidly. According to the consensus report of a major 2005 study, which uses the term "human germline genetic modification" or HGGM, advances in research reported in 2004 and 2005 have "overcome what were long regarded as impenetrable technical barriers, bringing the possibility of HGGM much closer. Therefore, the time is right for a new public discussion about whether, when, and how HGGM research should proceed."[4]

By one definition, human germline modification has already occurred. In 2001, a reproductive clinic in New Jersey reported success in "the first case of human germline genetic modification resulting in normal healthy children."[5] What they achieved, if it deserves to be called germline modification at all, was highly limited in its scope. Nevertheless, many observers agree that "the application of the rapidly emerging techniques of gene therapy to heritable human genetic modification is inevitable."[6]

Many technical difficulties must yet be overcome before germline modification can be regarded as acceptably safe for human use, and it is not clear when and how they will be overcome. There can be little doubt, however, that in a time frame and through developments we cannot foresee, some form of human germline genetic modification will become available in the not-too-distant future and that one day we will wake up to find ourselves overtaken by "the inevitability of new choices."[7]

If so, then a new discussion should begin before the technology is entirely in place. Anyone who has ever worried that morality too often lags behind technology might tolerate our being a little premature. The new discussion, broadened in its scope and the diversity of its participants, and drawing upon our collective human resources of moral and spiritual wisdom, should aim at creating the cultural resources necessary to illumine the human future, preferably before and not after the technology arrives on the scene. The advice of experts is clear: "[I]ndividuals and public advisory committees would be wise to begin the discussion of this important topic sooner rather than later."[8] The time has come to open up the discussion, to broaden its range of participants, and to bring to bear the moral and religious traditions that shape our values and our culture even today.

**Religion and Philosophy Have a Shared Role to Play in Public Debate**

Philosophical critics of human germline modification and reproductive cloning often point to religion as their partner in opposing these technologies. For instance, Francis Fukuyama and Leon Kass appeal to religious opposition to biotechnology to win support for their conclusions. They even praise religion, up to a point. Fukuyama says that religious objections to biotechnology are to be admired for their clarity and immediacy, for example, the "sharp distinction between human and nonhuman creation; [for] only human beings have a capacity for moral choice, free will, and faith, a capacity that gives them a higher moral status than the rest of animal creation."[9] Most of all, religion motivates or galvanizes resistance. As Fukuyama puts it, "religion provides only the most straightforward motive for opposing certain new technologies."[10] Furthermore, "religion often intuits moral truths that are shared by nonreligious people"[11]

Even so, for Fukuyama and Kass, the role of religion is limited. It may be a useful ally with great powers to mobilize public support, but theology is not appropriate for public argument. "While religion provides the most clear-cut grounds for opposing certain types of biotechnology, religious arguments will not be persuasive to many who do not accept religion's starting premises. We thus need to examine other, more secular, types of arguments."[12] Not wanting his own objections to germline modification to be dismissed as religion, Fukuyama seeks to separate his argument from religion. "I believe that it is important to be wary of certain innovations in biotechnology for reasons that have nothing to do with religion."[13]

According to Leon Kass, secular critics of biotechnology must take care to distinguish their own philosophical arguments from similar-sounding religious objections because philosophical or "serious moral objections . . . are often facilely dismissed as religious or sectarian."[14] Kass continues: "Religious thought—I would hesitate to call it theorizing—has its own profound understanding of the human condition and teachings about the moral life, an understanding deep enough to help us address the large questions of our humanity at stake in life's encounters with biotechnology. But the pluralistic premises of American ethical discourse and the fashions of the modern academy lead the mainstream to view such religious traditions at best with suspicion and often with outright contempt."[15] Philosophy should strip its arguments of "religious thought." However, should it fail to do so completely, then "never mind if these beliefs have a religious foundation—as if that should ever be a reason for dismissing them!"[16]

It is of course true that specific religious beliefs are not widely shared and may even be regarded with contempt or bewilderment by those outside a tradition. And it is true, as Fukuyama argues, that religious arguments are not likely to persuade the nonreligious. The same may surely be said of metaphysics, particularly the sort of metaphysical assertions about human nature employed by Fukuyama and Kass. If the contemporary secular academy dismisses religion, it is hardly hospitable to metaphysics. Outside the academy, the balance of popular support swings even more in the direction of religion. Of course, the validity of an argument does not depend at all upon the percentage of the population that finds it persuasive. However, the point made by Kass and Fukuyama is not that philosophy is true while religion is not, or even that philosophy's presuppositions are more universally plausible than those of any particular religion, but merely that philosophy is more popular in its persuasiveness than religion. This is an empirical claim that lacks support.

More damaging to the philosopher's case for the superiority of philosophy over religion in public debate is the fact that philosophers disagree among themselves. If disagreements among the religions count against religion having a public role, the same should be true of philosophy. This is especially obvious when we limit our scope to contemporary philosophers who have written on human germline modification. Along with Kass and Fukuyama, Jürgen Habermas has argued on philosophical grounds against such technologies as human germline modification. While agreeing in their conclusion that these technologies must be opposed, Habermas disagrees with Fukuyama and Kass on the basis for the opposition. Habermas in fact invokes the very argument that Fukuyama and Kass employ against religion and turns it into an argument against philosophical metaphysics, which is the foundation upon which Fukuyama, in particular, bases his argument.

Fukuyama argues that germline modification would violate human nature, which "is the sum of the behavior and characteristics that are typical of the human species, arising from genetic rather than environmental factors."[17] Then he asks: "What is it that we want to protect from any future advances in biotechnology? The answer is, we want to protect the full range of our complex, evolved natures against attempts at self-modification. We do not want to disrupt either the unity or the continuity of human nature, and thereby the human rights that are based on it."[18]

Habermas agrees with Fukuyama that human germline modification is wrong, but he rejects Fukuyama's line of argument as indistinguishable from religion. Philosophy must turn away equally from religion and metaphysics. Habermas warns against relying on "the classical image of humanity derived from religion and metaphysics."[19] Modern science has undermined confidence in metaphysics and religion equally. Human "nature" is conceptually adrift and technologically plastic. As much as he might want to restrain "technical self-optimization" by appealing to the classical views of a normative human nature, religious or metaphysical, Habermas warns against such a move. "Unless we fall back on treacherous metaphysical certainties, it is reasonable to expect persisting disagreements in the discourse universe of competing approaches to a species ethics."[20]

Our point is not to disparage philosophy or metaphysics as a public voice, or even to ask philosophers not to disparage religion while excluding themselves, but to suggest that in both cases, our powers to communicate and to persuade are limited. If so, then perhaps the right question to ask is this: What do we hope philosophy and religion will contribute to the public debate on questions like germline modification? If we hope for arguments that persuade majorities or unify cultures or justify legislation, we are likely to be disappointed. Such is not the role of religion or philosophy in today's context. But if we expect to deepen the debate, to enrich our understanding, and to pause long enough in our head-long rush to the future to draw upon traditional sources of human wisdom and well-tested accounts of human virtue, and if we hope to argue with fresh vigor while respecting deeply held differences, then metaphysics and religion may both have something to say.

Even today, many still find that religion has unique capacities to nurture in us that which is compassionate and devoted to the healing of others for the sake of nothing more than the healing of others, to lead us beyond a focus on ourselves while at the same time heightening our awareness of our susceptibility to the old temptations to which technology can add unexpected allure. Religion invites us to reflect on our weaknesses and anxieties so that we might know ourselves well enough to avoid some of the exploitations and high-tech seductions that might otherwise prey upon our fears, making sophisticated fools of us. Taken seriously, religion reminds us daily to do justice, to guard against new forms of discrimination and unfairness that might come from expansive powers, and to seek broad access to the benefits of technologically advanced medicine. All these things religion does in individual lives and in communities of faith, and in so doing affects the broader culture, adding to its collective wisdom, maturity, and depth.

## Correcting the Record: Religious Support for Human Germline Modification

It is widely believed that religious scholars and leaders oppose human germline modification, if not unanimously, then at least by a wide margin. Kass and Fukuyama assume this when they point to religion as support for their own objections. This view, however, is mistaken, and one of the more important contributions of this book is to set the public record straight. Religious support for germline modification is qualified and conditional, of course, but the majority of religious voices and nearly all the official statements of religious bodies leave the door open on the question of the morality of genetic modification of human offspring.

Why is it so often thought that religion is opposed to germline modification? One reason might lie in the public's tendency to exaggerate greatly the amount of conflict between science and religion. While historians of science have long since rejected the idea of warfare between science and religion, the news media and the general public still believe that these two arenas of human life are locked into some perpetual state of conflict. More often than not, religious scholars and institutions are supportive of science and technology, especially medicine, complaining only of the scientism that sometimes passes for science or specific methods of research, such as experiments involving human embryos.

More than any other, one phrase summarizes the warfare view, especially in the context of biomedical research and in such areas as human germline modification. That phrase is "playing God," which is most often used as a kind of verbal protest when it is felt that someone is going too far in making life-and-death decisions for other human beings. In that respect, the phrase resonates well in a secular society that defends autonomy, for the person who plays God intrudes not on God's sovereignty, but on the sovereign autonomy of another person. One of the classic uses of the phrase is found in the writings of a Protestant theologian, Paul Ramsey, who in the early 1970s wrote in opposition to the development of in vitro fertilization (IVF) techniques for human beings. According to Ramsey, "Men ought not to play God before they learn to be men, and after they have learned to be men they will not play God."[21]

This phrase has taken on a life of its own and is echoed today by many who share the idea that there must be limits to the use of biomedical technology, even by those whose objections are not based in religion. For example, Leon Kass uses the phrase this way: "By it is meant one or more of the following: man, or *some* men, are becoming creators of life, and indeed, of individual living human beings (*in vitro* fertilization, cloning); they stand in judgment of each being's worthiness to live or die (genetic screening and abortion)—not on moral grounds, as is said of God's judgment, but on somatic and genetic ones; they also hold out the promise of salvation from our genetic sins and defects (gene therapy and genetic engineering)."[22] Jürgen Habermas uses the phrase this way: "'Partner in evolution' or even 'playing God' are the metaphors for an auto-transformation of the species which it seems will soon be within reach."[23] In both cases, these philosophers use this phrase as a kind of rhetorical shorthand to warn that certain technologies go too far and that God (or those at least who believe in God) are opposed.

The myth of the warfare and the rhetoric of playing God all came together in the public theater in 1983 when a large and diverse group of religious leaders, such as Catholic bishops and Protestant denominational leaders, signed a highly publicized statement in opposition to germline modification. As much as anything, this event has created the impression that religious scholars and leaders are united in opposition to this technology. The statement was developed and promoted, not by a theologian or church leader, but by Jeremy Rifkin, an economist whose book *Algeny* came out the same month. Rifkin was able to secure the signatures of leaders from across the spectrum of religious bodies, including the most conservative and liberal Protestants, who sometimes signed without seeking scientific or theological advice. The document, which is worded as a resolution, comes to this conclusion: "[E]fforts to engineer specific genetic traits into the human germline should not be attempted."[24]

The fact that this statement was signed by many church leaders, such as bishops and heads of denominations, certainly lends support to the claim that Christian leaders are all opposed to germline modification. A closer examination of the official texts of the religious communities, however, leads to quite a different conclusion. The next section of this chapter reviews some of these texts. First, however, it is instructive to return for a moment to Paul Ramsey, who warned about the dangers of playing God. In the same book from which that quotation is taken, Ramsey endorses the idea of human germline modification. Already in the early 1970s he was able to foresee the possibility of what might lie ahead and far from condemning it, he strongly endorsed it: "The notation to be made concerning genetic surgery, or the introduction of some anti-mutagent chemical intermediary, which will eliminate a genetic defect before it can be passed on through reproduction, is simple. Should the practice of such medical genetics become feasible at some time in the future, it will raise no moral questions at all—or at least not that are not already present in the practice of medicine generally. Morally, genetic medicine enabling a man and a woman to engender a child without some defective gene they carry would seem to be as permissible as treatment to cure infertility when one of the partners bears this defect."[25] While Ramsey is wary of the possibility of playing God, he does not include human germline modification or genetic surgery under the heading of the prohibited.

Contrary to popular opinion, religious scholars and leaders are not unanimously opposed, but are in fact generally open to the possibility of a morally acceptable approach to human germline modification. Ever since the idea of genetic surgery was first discussed in the 1960s, some theologians and religious ethicists have recognized that germline modification may be technologically farfetched, but it is not obviously immoral or irreligious. In fact, precisely because of their religious convictions, many religious leaders and scholars over the past few decades have seen the idea of germline modification as morally preferable to any other response to the problem posed by the genetic transmission of disease. Germline modification, for all the challenges it poses, does offer some hope that an embryo may be treated rather than discarded, or that a healthy embryo might be created in the first place, and for such reasons it invites religious consideration by many. According to many religious scholars and leaders, including most of the contributors to this volume, germline modification is not obviously wrong but quite possibly is acceptable under certain conditions. This perspective is clearly present in the official statements of religious leaders and institutions, the subject of the next section of this chapter.

## Religion and Germline Modification: Cautious, Conditional Approval

Despite its public visibility, the 1983 letter is unique among religious statements, not just in the widespread but unreflective process that produced it or in its simplistic and categorical judgments, but mainly in its content. The letter refuses to leave the door open at all to the moral permissibility of human germline modification. If the letter is unusual, a more typical statement is found in the publications of the World Council of Churches (WCC), whose participant churches have a combined membership of over half a billion and include most Protestant denominations and Orthodox churches. After much study and review by the member denominations, the WCC issued a report in 1989 saying that "The World Council of Churches proposes a ban on experiments involving genetic engineering of the human germline at the present time, and encourages the ethical reflection necessary for developing future guidelines in this area."[26] For anyone reading too quickly, the word "ban" jumps out, confirming any prior notion that religion opposes germline modification. Read more carefully, however, the report clearly bases its opposition on safety grounds "at the present time" rather than on permanent moral grounds.

A similar position was endorsed in 2006 by the National Council of the Churches of Christ, USA, whose members include most U.S. Protestant denominations. In its report, the council states: "Effective germ line therapy could offer tremendous potential for eliminating genetic disease, but it would raise difficult distinctions about 'normal' human conditions that could support discrimination against people with disabilities. But the human community has some time to reflect on this conundrum. Inaccuracies in somatic gene therapy have resulted in activating dangerous nearby genes and led U.S. regulators to temporarily suspend all human gene therapy using viral vectors. As a result, the case for germ line therapy, which would affect not only those presently treated but all their descendants as well, has become even more difficult to make."[27] The statement carefully notes the advantages but also the social and moral challenges posed by the prospect of germline modification. It refers to current difficulties in gene therapy for human somatic cells, suggesting that these hurdles raise even more difficult challenges to safety that must be met before germline modification could ever be seriously entertained. The report does not, however, oppose the idea of germline modification, provided these concerns can be addressed.

The United Methodist Church, one of the largest Protestant denominations in the United States, has developed a comprehensive position on genetics. Generally speaking, the position is cautious, even restrictive. In 1992, the Methodist Church endorsed this statement of opposition: "Because its long-term effects are uncertain, we oppose genetic therapy that results in changes that can be passed to offspring (germ-line therapy)."[28] This wording is modified slightly in 2000:

> We oppose human germ-line therapies (those that result in changes that can be passed to offspring) because of the possibility of unintended consequences and of abuse. With current technology it is not possible to know if artificially introduced genes will have unexpected or delayed long-term effects not identifiable until the genes have been dispersed in the population.
>
> We oppose both somatic and germ-line therapies when they are used for eugenic purposes or enhancements, that is, to provide only cosmetic change or to provide social advantage.[29]

In the 2000 statement, opposition to germline modification is qualified by the reference to "current technology," which might change the moral assessment. In that reading of the statement, the core idea of germline modification for therapy is not opposed if safety can be assured in the long term and enhancement is avoided.

Perhaps more surprising is that the largest and most conservative major U.S. Protestant denomination, the Southern Baptist Convention, has left the door open to human germline modification, provided of course that safety concerns are resolved. In June 2006, the national gathering of the convention adopted a resolution aimed largely at restating objections to embryo research and to research that involves human–animal chimeras or mosaics. The resolution says this about germline modification: "RESOLVED, That we cannot endorse any use of human germline modification at this time, no matter how well-intentioned, due to the unpredictability of the process and the possible introduction of irreversible destructive errors into the human gene pool."[30] Here again, those who wrote and supported this wording were careful to base their objections on grounds of safety "at this time," thereby leaving open the door to reconsideration on moral grounds. While these statements cannot be read as endorsements of germline modification, they must be seen for their care not to endorse what cannot be done, but to leave the door open for now in tacit recognition that there are serious moral reasons in favor of germline modification.

Some might think that even though the Protestants have failed to condemn human germline modification, Catholics are surely reliable in making the religious case against any alteration of the human germline. Precisely the opposite is the case, for if anything, the Catholic statements more clearly define the good that might be gained by a germline approach. In the chapters that follow, James Walters and Thomas Shannon carefully show how Catholic theology does not lead to a categorical rejection of germline modification. On the contrary, as long as certain constraints are in place, the core idea of human germline modification is acceptable. One of these constraints—shared with some of the Protestant statements, such as the United Methodist position—is that germline modification must be for therapy only and avoid what might be called human enhancement. In addition, however, Catholic moral theology objects to human in vitro fertilization and the use of a human embryo for nontherapeutic purposes. In other words, it is not acceptable to create or to treat the embryo outside the human body, nor can one embryo be used to create or treat another embryo.

These constraints place strong but not insurmountable limits on germline modification. According to a high-level Vatican theological committee, human germline modification remains a possibility: "Germ line genetic engineering with a therapeutic goal in man would in itself be acceptable were it not for the fact that is it is hard to imagine how this could be achieved without disproportionate risks especially in the first experimental stage, such as the huge loss of embryos and the incidence of mishaps, and without the use of reproductive techniques. A possible alternative would be the use of gene therapy in the stem cells that produce a man's sperm, whereby he can beget healthy offspring with his own seed by means of the conjugal act."[31] Next to this statement, the comment of Pope John Paul II might be noted: "A strictly therapeutic intervention whose explicit objective is the healing of various maladies such as those stemming from chromosomal defects will, in principle, be considered desirable, provided it is directed to the true promotion of the personal well-being of the individual."[32]

The Vatican encyclical *Donum vitae* quotes these words of Pope John Paul II, offering its own statement in greater detail: "As with all medical interventions on patients, one must uphold as licit procedures carried out on the human embryo which respect the life and integrity of the embryo and do not involve disproportionate risks for it but are directed towards its healing, the improvement of its condition of health, or its individual survival."[33] While such procedures might not involve any germline modification, it is also clear from the context that they may do so, at least inadvertently. Together, these statements can be taken as reflecting the official teaching of the Catholic Church.

These statements should not be interpreted as endorsement for attempts at germline modification that ignore the constraints. The use of human germline modification for therapeutic purposes is a good and noble end, but it must not be pursued by means or techniques that violate the constraints. It must be noted, further, that honoring the constraints might mean that germline modification is never possible in a way that is morally acceptable. In the future, it might even turn out that human germline modification becomes possible in ways that satisfy the prevailing secular standard of safety, but nevertheless in a way that does not meet these Catholic standards and therefore is condemned by the church, but not because it is intrinsically wrong. Intrinsically, human germline modification for therapeutic reasons is morally acceptable to the Catholic Church.

From these statements, Protestant and Catholic, it may be concluded that the Christian churches generally do not oppose the core idea of human germline modification for therapeutic purposes. Two conditions have been noted in these statements. The first condition, shared implicitly if not explicitly by all the statements, is that any use of germline modification must be for therapeutic rather than enhancement purposes. In asserting this condition, no one is claiming to know precisely how to distinguish therapy from enhancement. However, in the most general terms, there is believed to be a difference between using this technology to allow the conception and birth of a child while diminishing the likelihood of a serious genetic disease, and using the technology to produce a child with socially desirable traits. The second condition, which is limited to the Catholic statements (although individual Protestants and Orthodox might agree), is that the means employed in human germline modification must avoid reproductive technologies such as in vitro fertilization or any nontherapeutic use of human embryos. An embryo may not be made to exist outside the body nor treated in a way that is not intended for its own benefit in respect to its developmental potential.

Of these two conditions, the first is generally endorsed by the contributors to this volume and by other scholars whose views are briefly noted in chapter 9. In chapter 5, Cameron and DeBaets make the important argument that the first condition needs to have the same sort of teeth as the second. If Catholic approval is to be withheld if the objections to IVF cannot be met, should not all (or nearly all) religious approval be withheld if the condition regarding therapy versus enhancement cannot be met? Cameron and DeBaets predict that even if the line between therapy and enhancement can be drawn, it cannot be held, for it is not in our nature to observe such a constraint. If approval is conditional upon observing a line between therapy and enhancement, and we expect that the line cannot be held, must religious scholars and leaders withhold approval? Chapter 9 will return to this question. In addition, it will explore another moral condition that must be met if germline modification is to be morally acceptable: Its use must be consistent with religious principles of social and economic justice. The justice condition is often noted but rarely developed in a thorough way, except by a few individual religious scholars.

In the chapters that follow, scholars in Judaism and Christianity reflect on the internal dynamics of their faith, which like Ramsey's thought is always more complex and subtle than the public recognizes. The authors focus on the question of germline modification by engaging it, not simply with a view to a yes or no answer, but as a context for a rigorous exercise in theological self-examination. Stated negatively, the goal is to counter the public view that religion is simplistic and monolithic, capable of little more than neo-Luddite complaints against modernity and technology. Positively, the goal is to open up some of the complexity of religious and theological reflection for the public in order to provoke a deeper discussion. In the next section of this chapter, however, attention is directed to the question of how to define human germline modification and what techniques might make it possible.

## Human Germline Modification—Definitions and Techniques

### Definitions

Human germline modification goes by several names, such as "germline gene therapy" or "designer babies." The term used here is human germline genetic modification, sometimes shortened to germline modification. The word "therapy" is avoided because of its strongly positive connotations. Until a medical technique is proven to bring about healing, it must be regarded as experimental and should not be called therapy. Furthermore, calling it therapy disguises the fact that in the end germline modification might be used primarily not for therapy but for what might be called enhancement.

If the term "therapy" is prejudicial in favor of the technology, the term "designer babies" is rhetorically negative, prompting thoughts of fashion design or trendy engineering, perhaps implying that any use of germline modification is the equivalent of designing a child with just the right features and options. Of course, germline modification might be criticized for harboring a secret tendency in that direction, but the criticism must be argued, not presupposed in the choice of terms. In contrast to both "therapy" and "design," the term "modification" is more precise and rhetorically neutral.

Germline genetic modification is called "germline" because the modification could pass to future generations. It affects the so-called germline cells, modifying their DNA in ways that may be inherited by offspring. A major study completed in 2005 defined germline modification this way: "Human Germline Genetic Modification refers to techniques that would attempt to create a permanent inheritable (i.e. passed from one generation to the next) genetic change in offspring and future descendants by altering the genetic makeup of the human germline, meaning eggs, sperm, the cells that give rise to eggs and sperm, or early human embryos."[34]

Embryos are included in the list of germline cells because modifying the genes of the embryo, if done at the time of fertilization, will affect all the cells that come from the early embryo, which include the germline cells, specifically sperm, eggs (or oocytes), and their precursors. Conversely, if oocytes or sperm or their precursors are modified, any embryos they produce will also be modified. In any case, the key point is that any modification of the DNA of germ cells could be inherited by future generations.

Germline modification is typically distinguished from somatic cell gene modification, which is most commonly known as somatic cell gene therapy or simply as gene therapy. In 1990, the first somatic cell gene modification was attempted, and since then hundreds of experiments have been conducted involving thousands of patients. The goal typically is to treat a genetic disease by modifying the DNA that causes it. Results so far have been disappointing, with little success and a few well-publicized individual tragedies that were setbacks for the whole field. Human germline modification, by contrast, targets the germline cells.

**Techniques**

If germline cells are the target, what are the techniques that might be used to modify their DNA? What are the procedures and technologies that might actually change the genes in germline cells? A range of strategies has been proposed. At present, all of the suggested strategies have limitations or problems that stand in the way of their being attempted in acceptably safe experiments. Nevertheless, nearly all of them are being used, one way or another, in experiments with nonhuman animals or with human cell cultures.

The techniques fall into two basic categories. Strategies in the first group focus on adding new DNA, whereas these in the second group attempt to correct or replace the existing DNA with a another segment. In the first group at least four types of techniques are being developed.

**Viral Vectors** Viruses are naturally able to transport DNA into living cells and insert it into the chromosomes. Viral vectors are viruses that have been modified to keep them from causing disease. The DNA to be inserted into the cells is first inserted into the modified virus. Millions of copies are produced and allowed to enter the target cells. The hope is that the transported DNA will begin to function inside the targeted cells, ideally overriding a genetic disease. There are two major problems with viral vectors. First, the inserted DNA may not end up in the right location, in which case it might not work, or worse, it might interrupt a normal gene. Second, the old, disease-related DNA remains, and so does the viral DNA itself, which could cause problems in a developing embryo or later in life.

**Nonviral Vectors** To avoid at least the problem of the viral DNA, some researchers have developed nonviral techniques for inserting DNA into cells. One approach is to insert just the DNA strand itself into the cell by microinjection. Another is to package the DNA in a tiny capsule of fatty substance that can pass into the cell. These and other techniques avoid the insertion of viral DNA but have the other problems associated with viral vectors.

**Artificial Chromosomes** A completely different approach to adding DNA involves constructing what amounts to a small version of a chromosome. The DNA in the nucleus of cells is packed into chromosomes that duplicate themselves when the cell divides. Researchers have been able to create human artificial chromosomes, imitating the basic structure found in nature but containing just the DNA that researchers build into it. The thought is that an artificial chromosome might be inserted into an embryo at fertilization. The main advantage of artificial chromosomes is that they can carry twenty to thirty times as much DNA as the largest capacity viral vectors.[35] Large genes and indeed many genes can be built into an artificial chromosome and transferred as a unit into a living cell. If used in germline modification, however, the presence of these chromosomes might cause chromosomal abnormalities, a serious health concern. Quite likely an artificial chromosome would have to be removed in the distant future when a person with germline modification seeks to reproduce.

**Ooplasm Transfer** This approach, which is of narrow interest but important because it has already been used, was developed as a way to help avoid a rare set of diseases known as mitochondrial disorders. Most DNA is located in chromosomes but a tiny portion is found in small structures outside the cell nucleus. These structures, called mitochondria, are inherited only from one's mother. If a woman with a mitochondrial disorder wants to have children, she knows that they will all inherit her disorder. In order to avoid this while helping her have children with her own nuclear DNA, researchers have developed a way to transfer ooplasm, which contains the mitochondria, from a donor egg to the prospective mother's egg and then fertilize the modified egg.[36] The DNA of the resulting children (mitochondrial, not nuclear) is modified by technology, and so in a minimal way this procedure falls within the scope of the definition for germline modification.[37] It is not likely that this technique will be used widely, but it is historically significant as the first use of human germline modification.

In addition to adding DNA, it may be possible to replace or repair the DNA that is present in the germ cell. The advantage of replacement or repair is that the old DNA sequence is not left behind, possibly causing health problems in the future. Two approaches have been proposed.

**Gene Repair** DNA mutates or changes spontaneously in the human body. These mutations could lead to disease, including cancer, but fortunately the cells themselves correct these errors. It is possible to mimic this function by constructing short sequences of DNA and its companion, RNA, and packaging the sequence so that it can enter specific target cells. There the DNA–RNA sequence finds the mutation, binds to it, and forces it to change.[38] If this technique can be successfully developed, it still faces an important limit. It is capable of correcting only the tiniest amount of DNA. A few genetic diseases are caused by a one-base mutation, and these might be treated through this approach. Or it might be possible to use this technique to disable or "knock out" a gene that could be causing a disease. The major advantage of gene repair as a strategy for germline modification is that it leaves no unwanted DNA behind.[39]

**Gene Targeting** This strategy is also known by a more technical term, "homologous recombination," which uses a series of steps precisely to replace a mutated gene or an unwanted DNA sequence with a corrected gene. The process is too complicated to use directly on embryos or on eggs or sperm, but it might be possible to use it to modify cells that can be made to produce eggs or more likely, sperm. This approach will likely require an intermediary step involving human embryonic stem cells. It has been shown that gene targeting can be used to produce precise genetic modifications of human embryonic stem cells. These modified stem cells, multiplying in a dish, can then be selected by separating out those cells that have the correct modification from those that do not.[40] The next step is to use these genetically modified stem cells to produce the precursors of sperm or oocytes. Researchers have done this with mice.[41] If this technique can be applied to human beings, it may be possible to modify stem cells and from them produce eggs or sperm that carry the modification. From that point it would be relatively easy to create embryos with the genetic change.

One important question has to do with what is sometimes called inadvertent germline modification. When researchers are attempting gene modification of somatic cells, how do they know that they are not modifying the germline cells in the patient's body? They may be trying to avoid germ cells, but if they affect them, does this count as germline modification? The answer is yes, according to the definition used in 2000 by the American Association for the Advancement of Science (AAAS). According to this study, "inheritable genetic modifications [IGM] refer to the technologies, techniques, and interventions that are capable of modifying the set of genes that a subject has available to transmit to his or her offspring. IGM includes all interventions made early enough in embryonic or fetal development to have global effects on the gametes' precursor tissues, as well as the sperm and ova themselves. IGM encompasses inheritable modifications regardless of whether the intervention alters nuclear or extranuclear genomes, whether the intervention relies on molecular genetic or other technical strategies, and even whether the modification is a side effect or the central purpose of the intervention."[42] For the AAAS study, inadvertent modification of a germline was specifically included in the scope of the definition, primarily because of its link to research occurring today in somatic cell modification.[43]

The question of inadvertent modification of a germline is not directly addressed by the authors of the chapters that follow. Indirectly, however, the issue is always before us. Whether such modification is permissible may be the most important relevant public policy question on the immediate horizon. If inadvertent modification of a germline must be avoided without exception, then gene modification of somatic cells comes under a huge burden of proof that it is avoiding all germline changes. Such a policy might preclude certain techniques from ever being accepted.

What is under consideration here is not inadvertent bad effects. What is at stake is whether inadvertent beneficial effects might be permitted or whether, simply because they affect the germline, they must be banned regardless of their benefit. For example, if researchers treating a patient for a genetic disease eliminate the disease-linked DNA from germ cells, and if the patient then produces children who are free of the disease, have the researchers acted immorally and should public policy prevent them from doing so? Some might say yes if they believe that germline modification is inherently evil and that researchers have an obligation to avoid even a low degree of risk. Others, however, will say that under some circumstances, germline modification is not evil and that in such a case good has been done twice, first to the patient and then to the offspring.

To a large extent, our acceptance of inadvertent modification of a germline hinges on our moral stance toward intended germline modification. If so, then the debate over the morality of germline modification has immediate public policy implications, affecting how we regulate today's proposals for modification of somatic cells.

## Deepening the Discussion, Enriching the Debate

The contributors to this volume draw upon living religious traditions to widen and enrich the public debate over the human future. Elliot N. Dorff provides a helpful general introduction to the various ways religions draw upon ancient texts and traditions to make sense of contemporary challenges. He notes that from the tradition and perspectives of Judaism, there is a strong presumption in favor of medicine and the moral legitimacy of altering the natural world for a good purpose, and thus in favor of germline modification. At the same time he raises profound worries about human weakness and the ensuing potential for misuse of powerful technologies, and so he cautions us to proceed with care and with open deliberation.

Thomas A. Shannon clearly sets out the official teachings of the Catholic Church related to biomedical research in general and embryo research in particular. The core moral principle is that the dignity or value of human life must be protected without qualification from conception. A human embryo may be treated medically if the objective is therapeutic for the embryo. Germline modification, therefore, is morally acceptable. However, it is also true that in official teaching, the human embryo may never be used as a means to an end, whether to expand knowledge or to treat another person. This has implications for embryonic stem cell research and the use of nuclear transfer (cloning) for research purposes or as a way of treating another person. These constraints also limit the methods by which germline modification might be achieved, with the effect that what is permissible in principle might be impossible in practice. Shannon concludes with a review of his own criticisms of these constraints.

H. Tristram Engelhardt, Jr., notes the multiplicity of religions and even of Christianities, insisting upon the one he calls traditional. On the basis of traditional Christianity, he identifies specific limitations or conditions that must be met by germline modification. With these in place, he concludes on the basis of the core theological doctrines of traditional Christianity that a curative or therapeutic use of germline modification is permissible, possibly even obligatory. Perhaps most interesting is that Engelhardt's approval of germline modification is not limited to therapy in the usual sense but takes in a much wider scope, based on the distinctive features of traditional Christian doctrine. According to the traditional doctrine of creation, God creates human beings to be immortal, and while immortality is lost owing to the Fall, human longevity in Biblical times is ten times greater than it is today. There is no objection here to efforts to extend the human life span, a point that complicates any notion of a universally obvious breakpoint between therapy and enhancement. At the same time, any hope of a transhuman or posthuman future is seen as a poor substitute for the expected transformation that comes in the future of humanity divinized or made to participate in the life of the divine.

Nigel M. de S. Cameron and Amy Michelle DeBaets, reflecting the core anthropological insight of classical protestantism, insist that human nature is defined theologically in relation to God as its source (creator) and destiny (assumed in Christ). Technology must never aim to go beyond human nature as created or given by God and as assumed or taken up by God, as if we were permitted to transcend our natures by biomedical enhancement. This amounts to a categorical objection to enhancement. Yet this is exactly what germline modification will do, they argue, and any thought that its use can be limited to therapy is delusional. Therefore, the only religiously responsible position is to stop the whole field.

James J. Walter draws upon core Catholic doctrines to explore the question of germline modification, concluding that it is theologically legitimate. It is not, as some charge, an illicit act of playing God, as if it were an intrusion on God's sole prerogatives in respect to the creation of human life. In particular, he rejects the view that nature is fixed or static and that any technological modification violates natural order. On the other hand, we should be wary of our tendency to let technology become the raw assertion of human will over nature, as if no inherent goodness and purpose were present in nature to constrain our acts. Better to think of our actions as a kind of co-creation that honors God's purposes and contributes to them. The development of germline modification presents great challenges. Based on a theological analysis of major themes of the Catholic faith, however, Walter concludes that germline modification is defensible as consistent with a theological view of God's creative and redemptive purposes.

Lisa Sowle Cahill agrees that germline modification is acceptable for therapy but not for enhancement. Her essay raises questions about the difficult concept of human nature, which she identifies, not with a list of fixed properties grounded in the biology of each individual, but as arising from our sociality, which enables our flourishing when it is characterized by justice. The problem of enhancement is that it threatens to undermine justice and therefore poses a threat to our nature as social.

Continuing some of these themes, Celia Deane-Drummond focuses on the question of human moral agency and how traditional notions of conscience and virtue might apply to case-by-case uses of germline modification. On this basis she concludes that we should not rule out germline modification. In addition to attention to ourselves as moral agents, we also need to consider the methods that might be used to achieve germline modification. Instrumental or nontherapeutic use of embryos, for example, is not permissible, and so any strategy of germline modification that requires the creation and destruction of embryos is ruled out.

The final chapter explores more fully the religious case in favor of human germline modification, examining at length the moral conditions that are often tied to that approval. The chapter concludes with a return to the challenge posed by Hans Jonas, which focuses our attention not so much on the technologies of human transformation, but on those human beings who will use them and those who will be made different by them.

## Notes

1. Hans Jonas, *The Imperative of Responsibility: In Search of an Ethics for the Technological Age*, trans. Hans Jonas and David Herr (Chicago: University of Chicago Press, 1984), p. 21.

2. Susanna Baruch, Audrey Huang, Daryl Pritchard, Andrea Kalfoglou, Gail Javitt, Rick Borchelt, Joan Scott, and Kathy Hudson, *Human Germline Genetic*

*Modification: Issues and Options for Policymakers* (Washington, DC: Genetics and Public Policy Center, 2005), pp. 28–29.

3. A. W. S. Chan, K. Y. Chong, C. Martinovich, C. Simerly, and G. Schatten, "Transgenic monkeys produced by retroviral gene transfer into mature oocytes," *Science* 291 (January 12, 2001): 309–312.

4. Baruch et al., *Human Germline Genetic Modification*, p. 9.

5. J. A. Barritt, C. A. Brenner, H. H. Malter, and J. Cohen, "Mitochondria in human offspring derived from ooplasmic transplantation," *Human Reproduction* 16:3 (March 2001): 513–516 at 513. See also Erik Parens and Eric Juengst, "Inadvertently crossing the germ line," *Science* 292 (April 20, 2001): 397.

6. Theodore Friedmann, "Approaches to Gene Transfer to the Mammalian Germ Line," in Audrey R. Chapman and Mark S. Frankel, eds., *Designing our Descendants: The Promises and Perils of Genetic Modifications* (Baltimore: Johns Hopkins University Press, 2003), p. 39.

7. Allen Buchanan, Dan W. Brock, Norman Daniels, and Daniel Wikler, *From Chance to Choice: Genetics and Justice* (Cambridge: Cambridge University Press, 2000), p. 18.

8. LeRoy Walters and Julie Gage Palmer, *The Ethics of Human Gene Therapy* (New York: Oxford University Press, 1997), p. 78.

9. Francis Fukuyama, *Our Posthuman Future: Consequences of the Biotechnology Revolution* (New York: Farrar, Straus and Giroux, 2002), p. 88.

10. Fukuyama, *Posthuman Future*, p. 90.

11. Fukuyama, *Posthuman Future*, p. 90.

12. Fukuyama, *Posthuman Future*, p. 91.

13. Fukuyama, *Posthuman Future*, p. 12; cf. 161.

14. Leon R. Kass, *Life, Liberty and the Defense of Dignity: The Challenge for Bioethics* (San Francisco: Encounter Books, 2002), p. 7.

15. Kass, *Life, Liberty*, p. 60.

16. Kass, *Life, Liberty*, p. 114.

17. Fukuyama, *Posthuman Future*, p. 130.

18. Fukuyama, *Posthuman Future*, p. 172.

19. Habermas, *The Future of Human Nature* (Cambridge, UK: Polity Press, 2003) p. 93.

20. Habermas, *Future of Human Nature*, p. 93.

21. Paul Ramsey, *Fabricated Man: The Ethics of Genetic Control* (New Haven, CT: Yale University Press, 1970), p. 138.

22. Kass, *Life, Liberty*, p. 129; italics in original.

23. Habermas, *Future of Human Nature*, p. 21.

24. Rifkin, Jeremy, "The Theological Letter Concerning Moral Arguments Against Genetic Engineering of the Human Germline Cells" (Washington, DC:

Foundation on Economic Trends, 1983); cf. Ronald Cole-Turner, *The New Genesis: Theology and the Genetic Revolution* (Louisville, KY: Westminster John Knox Press, 1993), pp. 74–75; and John H. Evans, *Playing God? Human Genetic Engineering and the Rationalization of Public Bioethical Debate* (Chicago: University of Chicago Press, 2002), pp. 165–173.

25. Ramsey, *Fabricated Man*, p. 44.

26. The World Council of Churches, "Biotechnology—its challenges to the churches and the world: Report by the WCC Subunit on Church & Society," (Geneva: August 1989). Available at http://www.oikoumene.org/en/resources/documents/wcc-programmes/justice-diakonia-and-responsibility-for-creation/science-technology-ethics/08-89-biotechnology.html (accessed May 7, 2007).

27. National Council of Churches USA, "Fearfully and Wonderfully Made: A Policy on Human Biotechnologies" (New York, 2006). Available at http://www.ncccusa.org/pdfs/adoptedpolicy.pdf (accessed May 7, 2007).

28. United Methodist Church, *Book of Discipline of the United Methodist Church* (Nashville, TN: United Methodist Publishing House, 1992), pp. 97–98.

29. The United Methodist Church, *The Book of Resolutions of the United Methodist Church 2000* (Nashville, TN: United Methodist Church, 2000), paragraph 90, "New Developments in Genetic Science."

30. Southern Baptist Convention Annual Meeting, Resolution 7 "On Human-Species Altering Technologies" (June 2006). Available at http://www.sbcannualmeeting.org/sbc06/resolutions/sbcresolution-06.asp?ID=7 (accessed May 7, 2007).

31. International Theological Commission [Catholic Church], "Communion and Stewardship: Human Persons Created in the Image of God," paragraph 90 (Vatican, 2002). Available at http://www.vatican.va/roman_curia/congregations/cfaith/cti_documents/rc_con_cfaith_doc_20040723_communion-stewardship_en.html (accessed May 7, 2007). Published with the permission of Joseph Cardinal Ratzinger, then the president of the Commission and now Pope Benedict XVI.

32. John Paul II, "Dangers of Genetic Manipulation," Address to the World Medical Association, 1983.

33. The Holy See, *Donum vitae. Instruction from the Congregation of the Doctrine of the Faith,* Vatican, February 1987, I, 3.

34. Baruch, "Human Germline Genetic Modifications," p. 9.

35. Cf. Nobutaka Suzuki, Kazuhiro Nishii, Tuneko Okazaki, and Masashi Ikeno, "Human artificial chromosomes constructed using the bottom-up strategy are stably maintained in mitosis and efficiently transmissible to progeny mice," *Journal of Biological Chemistry* 281 (September 8, 2006): 26615–26623.

36. Barritt et al., "Mitochondria in human offspring."

37. See Parens and Juengst, "Inadvertently crossing the germ line."

38. See R. Michael Blaese, "Germ-Line Modification in Clinical Medicine: Is there a Case for Intentional or Unintended Germ-Line Changes?" in Chapman and Frankel, eds., *Designing our Descendants*, pp. 68–76.

39. See Blaese, "Germ-Line Modification" and Kenneth W. Culver, "Gene Repair, Genomics, and Human Germ-Line Modification," in Chapman and Frankel, eds., *Designing our Descendants*, pp. 77–92.

40. T. P. Zwaka and J. A. Thomson, "Homologous recombination in human embryonic stem cells," *Nature Biotechnology* (March 21, 2003) 21: 319–321.

41. H. Kubota, M. R. Avarbock, and R. L. Brinster, "Growth factors essential for self-renewal and expansion of mouse spermatogonial stem cells," *Proceedings of the National Academy of Sciences USA* (November 23, 2004) 101: 16489–16494.

42. See Mark S. Frankel and Audrey R. Chapman, *Human Inheritable Genetic Modifications: Assessing Scientific, Ethical, Religious, and Policy Issues* (Washington, DC: American Association for the Advancement of Science, 2000). Available at http://www.aaas.org/spp/sfrl/projects/germline/report.pdf.

43. E. Juengst and E. Parens, "Germ-line Dancing: Definitional Considerations for Policymakers," in Chapman and Frankel, eds., *Designing Our Descendants*, pp. 20–39.

# 2

# Judaism and Germline Modification

Elliot N. Dorff

## Perspectives and Methods

Why is it important to be aware of varying religious and secular perspectives on moral matters in the first place? Why, in other words, is it that morals do not come in one universal and eternal set of norms, but rather differ among religions, societies, and times?

The answer is embedded in the very word "religion." The "lig" in that word comes from the Latin root meaning "to tie together," the same root from which we get the word "ligament," which is connective tissue, and tubal "ligation," which is tying a woman's reproductive tubes to prevent pregnancy. Among other functions, religions describe our ties to our family, community, the whole human species, the environment, and the transcendent (imaged in the western religions as God). That is, religions give us a broad picture of who we are and who we ought to be. Secular philosophies (western liberalism, Marxism, existentialism, etc.) provide such perspectives as well. Indeed, what passes for secular ethics in western countries is rooted in the particular viewpoint of western liberalism, the product of such people as John Locke and Claude Montesquieu. While secular theories generally are produced by one person or a few people, attracting whoever becomes convinced of a particular philosophy, religions from their very origins are more likely to be tied to a group that endeavors to live out the religion's vision, using rituals, symbols, liturgy, and songs to remind adherents of that perspective and to induce continued loyalty to it. Furthermore, religious visions include attention to the transcendent element of human experience, while secular philosophies usually denigrate, ignore, or deny it altogether.

The various religions of the world, then, articulate their own particular views of how people are and ought to be. They each suggest a particular pair of eyeglasses, as it were, through which we should look at life. None of us can see the world without such lenses, for none of us is omniscient. We instead must perforce look at the world from our own vantage point; Einstein's theory of relativity applied not just to our knowledge of objects but also to our knowledge of everything else. So, for example, the Jewish, Christian, and western liberal lenses have much in common, but they also differ in significant ways.[1]

Specific moral norms are rooted in such big pictures. The differences in how the U.S. Supreme Court, Catholic doctrine, and Jewish law understand the status of the fetus, for example, leads to their differing positions on abortion and, in the case of Catholics and Jews, on embryonic stem cell research as well. Similarly, the differing ways in which those three systems of thought perceive the relationship between human beings and nature explain and motivate some of the ways in which they disagree with each other in understanding our place in life and what we may and should do with modern technology in altering our world.

Every tradition, whether religious or secular, has its own way of addressing questions posed to it. Catholicism, for example, invokes the authority of the Magisterium and, ultimately, the Pope to decide moral issues, even though moral theology and past decisions of popes and church councils also play a role in shaping current church doctrine. Protestants look to the Bible, personal conscience, and, to some extent, the traditions of their specific denomination to decide moral matters. American secular thought appeals to pragmatism and individual freedom to determine which moral issues should be addressed communally in the first place. Then, if it is decided that a communal response is appropriate, majority vote is the primary and fundamental method of making decisions, although that is modified by representative government, appointed officials, and constitutional concerns. To address moral questions, Judaism uses a variety of materials together with the methods suitable to them. These include stories, proverbs, theology, and historical experience, but Judaism puts primary emphasis on case-based law.[2]

No matter what method a particular tradition uses, however, it must stretch to address many of the new questions posed by modern medical technology and the context of modern medical health care. How traditions do that varies not only from one tradition to another but even within a particular tradition. Thus some Americans think that legislative action is the best way to address new medical questions; others think that the courts should treat these cases by developing case law sensitive to the intricacies of particular situations; some think that the government should stay out of these affairs altogether, leaving as much as possible to individuals to decide; and still others think that the appropriate method depends on the issue. Even Catholics, who seemingly have a clear, regimented system to decide moral matters, differ among themselves as to the extent to which the Vatican should determine such issues, with some claiming that individual Catholic moral theologians and even the individual Catholic parishoner should have a greater role in formulating Catholic responses to modern issues.[3]

Without a central authority like the Pope, Congress, or the Supreme Court, Jews differ among themselves even more markedly than Catholics or Americans do as to how to gain moral guidance from their tradition. Orthodox Jews look for precedents in established law, claiming that in that law God provided the answers to even the most contemporary issues if we would only be clever and persistent enough to interpret the law correctly. That does not prevent Orthodox Jewish writers from disagreeing with each other; on the contrary, some of the most vehement disputes occur within Orthodox circles. All of them, however, presume that the right answer, the one God wants us to reach, is contained wholly within the received tradition, if only we apply it rightly.

Conservative Jews—and in all fairness the reader should know that I am a Conservative rabbi—also believe that Jewish law should be used to give authoritative directions to our moral quandaries, but Conservative ideology asserts that from its very beginning Jewish law was the product of God and human beings, and it must be so today as well. Furthermore, the law appropriately changed over time to fit new circumstances, and it must be open to such changes in our own time as well. It is the rabbis of each generation, then, who must be entrusted with interpreting and applying the law to modern circumstances, but they cannot do that mechanically because Jewish law is silent about many modern issues that did not exist in the past. In stretching Jewish precedents to apply to those issues, then, rabbis must keep in mind past as well as new Jewish understandings of God and what God wants of us as embedded in Jewish law, thought, stories, and proverbs; current historical and economic circumstances relevant to such decisions; and the needs and moral sensitivities of contemporary Jews.

Reform Jews focus on individual autonomy. Thus while rabbis, physicians, and others may and should help individual Jews in making their medical decisions, and while Jews need to know the Jewish tradition in order to make a recognizably Jewish decision, ultimately individuals may and should decide for themselves how they are going to respond to all issues in their lives. Judaism may influence their decision, but each individual must decide whether it will and, if so, how.[4]

## Some Germane Principles Embedded in the Jewish Lens

To give readers an idea of both the content and the methodologies of Jewish medical ethics that are relevant to Judaism's understanding of germline modification, readers need to be familiar with several core Jewish beliefs that inform the Jewish discussion of all medical issues. In my book on Jewish medical ethics, I describe other central convictions pertinent to other aspects of health care,[5] but these will suffice for our purposes in this chapter:

### God Owns the Whole World, Including Our Bodies

The Torah (the first five books of the Bible) proclaims, "Mark, the heavens to their uttermost reaches belong to the Lord your God, the earth and all that is on it;" and Psalms declares, "The earth is the Lord's and all that it holds, the world and its inhabitants."[6] We human beings can only own property vis-á-vis other human beings, but God ultimately owns everything. Thus God can and does demand that we use some of "our" property to worship Him and to provide for poor people, for in the end our property is His.[7]

This includes ourselves. The Talmud, which is a record of rabbinic discussions of Jewish ethics, maintains that there are three partners in the creation of each one of us—mother, father, and God. Unlike our parents, however, God owns everything He created, including our bodies.[8] As a result, we have a responsibility to God to safeguard our health and life,[9] and, conversely, to avoid danger and injury.[10]

This appears to be in sharp contrast to the secular American point of view, which permits me to engage in downright dangerous behavior, refuse all medical treatment, or even commit suicide,[11] although not to assist someone in committing suicide.[12] All of this stems from the American presumption that my body is, after all, mine.

This, though, overstates matters, for even on secular American grounds, I must take into account not only the direct, physical effects of my actions on others, but indirect consequences as well. So, for example, American law requires me to wear a seatbelt while driving or riding in a car, presumably because my failure to do so might raise other people's insurance premiums or threaten their physical or economic welfare if we were involved in an accident. For reasons that are less clear, while American law permits the sale and use of tobacco, which is known to be addictive and carcinogenic not only to the smoker but also to those who inhale secondhand smoke, it nevertheless prohibits and punishes the use and, even more, the sale of marijuana, even if users take steps to ensure that they do not engage in behavior that might endanger others, like driving, and even if the marijuana is used to alleviate pain in a dying patient. Thus one cannot simply assert that American law permits me to do with my body anything I wish as long as I do not harm you. Still, people rightfully presume much more authority to determine what they may do with their bodies if they think that they are dealing with their own property rather than God's.

**Each Person Is an Integrated Whole, Both Personally and Communally**

While we can certainly think and talk about our various faculties separately, the Jewish tradition insists that we are integrated wholes. That is, the body, mind, emotions, and will all interact and affect each other. Thus the rabbis tell a story in which first the body and then the soul wants to deny responsibility for wrongdoing, but God "throws" the soul into the body and says, "This is how you were created, and this is how you will be judged."[13] Conversely, the proper path for a person in life is to cultivate both the body and the soul, for focusing on one to the neglect of the other leads one to sin. Thus the rabbis say, "Study of the Torah is beautiful (commendable) when combined with a gainful occupation, for when a person toils in both, sin is driven out of the mind. Study of Torah without work leads to idleness and ultimately to sin."[14] Thus health care in general and the effects of germline modification in particular must be considered on multidimensional levels, attending to the whole person and not to his or her body alone.[15]

Furthermore, while in the American law and thought Americans are individuals with "inalienable rights . . . of life, liberty, and the pursuit of happiness," the Jewish tradition portrays Jews as members of a community that God rescued from Egypt and brought to Mount Sinai, there to get not a single right but rather 613 duties. As a result, in America all communities are voluntary, which one can join or leave at will; even American citizenship, while difficult to obtain, can be renounced easily. In Jewish law, in contrast, once a person has been born to a Jewish woman or has converted to Judaism as an adult, he or she can never cease being Jewish for each Jew is organically connected to every other Jew, just as the parts of the body are tied to each other. A Jew who converts to a another religion becomes an apostate, losing all the privileges of being Jewish, but he or she still has all the obligations.

This tight bond to the community has immediate and strong implications for each Jew's responsibilities for the welfare of others. This includes not only providing food, clothing, and housing for the poor and education for everyone, but also thinking carefully about the social implications of any proposed public action or policy, including something very new like genetic modification. As the Talmud says, "Whoever is able to protest against the wrongdoings of his family and fails to do so is punished for the family's wrongdoings. Whoever is able to protest against the wrongdoings of his fellow citizens and does not do so is punished for the wrongdoings of the people of his city. Whoever is able to protest against the wrongdoings of the world and does not do so is punished for the wrongdoings of the world."[16]

**Medicine Is a Good Thing**

While some biblical passages assert that God governs illness and health,[17] the rabbis found the justification for human beings to engage in medical care based on other passages.[18] Rabbi Joseph Karo (1488–1575), author of the *Shulhan Arukh*, an authoritative code of Jewish law, goes further: he maintains that a physician who fails to try to heal when he can is effectively a murderer.[19] Furthermore, Jews may not live in a city with no physician,[20] for then they could not get the expert help they need to fulfill their responsibility to God to take care of their bodies. Physicians, in fact, have been very much honored in the tradition, and until the last century, when medical education began to require a decade or more of training, many rabbis also served as physicians and engaged in medical research.[21] God ultimately controls illness and health, but the physician is God's agent and partner in the ongoing act of healing: "Just as if one does not weed, fertilize, and plow, the trees will not produce fruit, and if fruit is produced but is not watered or fertilized, it will not live but die, so with regard to the body. Drugs and medicaments are the fertilizer, and the physician is the tiller of the soil."[22]

The talmudic image of human beings as God's partners in creation appears in B. *Shabbat* 10a and 119b. In the first of those passages, it is the judge who judges justly who is called God's partner; in the second, it is anyone who recites Genesis 2:1–3 (about God resting on the seventh day) on Friday night who thereby participates in God's ongoing act of creation. The Talmud in B. *Sanhedrin* 38a specifically wanted the Sadducees not to be able to say that angels or any being other than humans participate with God in creation.

This positive view of medicine, combined with Judaism's organic sense of community, has an important implication for health care generally and for germline modification in particular; namely, that the community as a whole has a duty to provide requisite health care to everyone, including the support of research to overcome illness and disability. This duty is based on the biblical passages "Do not stand idly by the blood of your brother" (Leviticus 19:16) and "Love your neighbor as yourself" (Leviticus 19:18). The Talmud uses the former verse to establish a positive duty to come to the aid of others: "On what basis do we know that if a man sees his fellow drowning, mauled by beasts, or attacked by robbers, he is bound to save him? From the verse, 'Do not stand idly by the blood of your neighbor.'"[23] Furthermore, the Talmud and Rabbi Moses ben Nahman (Nahmanides, 1194–1270) argue that "Love your neighbor as yourself" gives an express warrant to try to bring cure even when that involves the infliction of wounds through surgery or other risks to the patient, for everyone would (or should) prefer such risks to certain death. They also argue that the same verse also requires us to spend money to heal others if we lack the expertise.[24] (In the Jewish tradition, the biblical command to "love your neighbor as yourself" is understood to require, not only the feelings of caring for others, but specific behaviors that manifest that attitude, of which the provision of health care is one.)[25] To the extent that gene modification promises to provide or restore health, then, and to the extent that its cures can be made readily available to everyone who needs them, these aspects of the Jewish tradition would encourage it, allotting funds and energy to this form of research by weighing its probability of achieving such results in contrast to other promising therapies.

**Jews Have a Duty to God to Fix the World and to Preserve It**

What about gene modification to enhance human life? A Jew's duty to act responsibly is not limited to refraining from harming others, it includes also the obligation to work toward fulfilling the Jewish mission of *tikkun olam*, fixing the world. As the Mishnaic tractate, *Ethics of the Fathers*, asserts: "The world rests on three things—on Torah, on service of God, and on deeds of love (*gemilut hasadim*)."[26] The last of these is the Hebrew term used in earlier times for what we now call also *tikkun olam*.

This emphasis of classical Judaism continues to our own day and characterizes even Jews who are otherwise not very religious in their beliefs or practices. Modern Jews, in fact, often think of *tikkun olam* as *the* core commitment of Judaism. Thus in a 1988 national poll of American Jews conducted by the *Los Angeles Times*,[27] fully half listed a commitment to social equality as the most important factor in their Jewish identity.

Does this extend to fixing the world genetically? That depends, in part, on Judaism's understanding of technology. Adam and Eve are told in the Garden of Eden "to work it *and* to preserve it" (Genesis 2:15). Judaism has ever since tried to strike a balance between using the world for human purposes while still safeguarding and sustaining it. We are not supposed to desist from changing the world altogether: "Six days shall you do your work" is as much a commandment as "and on the seventh day you shall rest (literally, desist)" (Exodus 23:12).

In changing the world to accomplish our ends, though, we must take care to preserve the environment. That is true whether we are practicing medicine, farming, traveling, or doing anything else. Thus, although the Psalmist asserts that "The heavens belong to the Lord, but the earth He gave over to human beings,"[28] the rabbis make it clear that people may not use this divine gift wantonly, but rather must take care to preserve it: "Observe the work of God, for who can repair what he has ruined? At the time that the Holy One, blessed be He, created the first man, he took him around and showed him all its trees of the Garden of Eden. He said to him: 'Observe my creations, how beautiful and praiseworthy they are. Everything I created, I created for you. Take care not to ruin or destroy my world, for if you ruin it, there is nobody to fix it after you'."[29]

While we clearly must preserve the world, then, during the duration of our life, we may and should act as God's agents to improve it. God in fact intended that we function in that way. This is probably most starkly stated in a rabbinic comment about, of all things, circumcision. If God wanted all Jewish boys circumcised, the rabbis ask, why did He not create them that way? The answer, according to the rabbis, is that God deliberately created the world in need of fixing so that human beings would have a divinely ordained task in life, thus giving human life purpose and meaning.[30] We are then not only permitted, but mandated to find ways to bend God's world to God's purposes and ours—as long, again, as we preserve God's world in the process.

Just because we can do something, however, does not automatically mean that we ought to do it. To determine whether we should, we must measure its effects against Judaism's broader picture of our own good and that of God's world.

Thus technology in and of itself is not good or bad, it depends upon how we use it. If we employ it to assist us in shaping the world to achieve morally good ends while yet preserving the world, our use of technology is theologically approved and morally good. If, on the other hand, we disregard our duty to preserve the world when using technological tools, we are engaged in a theologically and morally bad act. Contrary to natural law theory, though, Judaism does not presume that the world that God created is ideal in its present state and therefore the standard by which we should judge every potential human intervention. On the contrary, God created the world to be fixed, and we humans need to determine when and how to aid God in that process.

While the Jewish tradition mandated fixing the world medically, legally, socially, and economically, it did not know about the potential for genetic enhancement. In weighing that prospect from a Jewish perspective, one must take note of the strong Jewish mission to fix the world, a mission that runs deep in the heart of contemporary Jews, even those who are not otherwise very religious. At the same time one must take account of God's ownership of the world and our duty to God to keep it from harm. Formulating a proper Jewish assessment of germline modification thus requires us to balance these duties, and this chapter is one attempt to do that.

## Five Jewish Texts Relevant to Modifying Human Nature

How shall we apply these Jewish methods and principles to the prospect of modifying human genetic structure, not only in one individual through his or her somatic cells but in that person's germline and thus his or her future generations as well? I would like to explore five Jewish texts that can guide us. Unless one accepts the Orthodox Jewish claim that the Torah was written by God, the authors of none of these texts knew about germline modification, much less intended to regulate our use of it, but each of these texts can, I think, shed significant light on how the Jewish tradition can be applied to that issue. Limitations of space prevent me from discussing any of these texts in detail, but what follows will, I hope, be enough to ground some important features of what I will formulate as at least one possible Jewish understanding of germline modification.

### The Tower of Babel

Chapter 11 of Genesis tells a story that is clearly intended as an etiological myth; that is, a story that explains the origins of a phenomenon, in this case why people speak different languages. In the process, though, we learn about theologically imposed limits to human industriousness. What is it that God finds objectionable in that story so that he confuses the workers' communication so that they cannot finish their project?

While the story can certainly be interpreted in a variety of ways, the rabbis of the Talmud understood the problem to be a lack of humility, which probably is the plain meaning of the story. Human beings wanted a tower to reach to the heavens "to make a name for ourselves; else we shall be scattered all over the earth."[31] Exactly how they thought that building a tower would give them a reputation that would prevent their being scattered is not clear, but the rabbis blame them for the hubris involved in thinking that they could reach the heavens and presumably become as powerful as God. In contrast, the Talmud depicts God as saying that Israel is beloved by God for being humble: "The Holy One, blessed be He, said to the People Israel: I love you because even when I bring you greatness, you deprecate yourselves to Me; but the other nations of the world, idolaters, are not like that: I gave greatness to Nimrod, and he said, 'Let us build ourselves a city [and a tower with its top in the sky]'" (Genesis 11:4).[32]

Where did Israel deprecate itself despite being granted greatness by God? The commentators list these examples: Abraham claiming that he is "but dust and ashes" (Genesis 18:27); Moses and Aaron saying, "And what are we?" (Exodus 16:7–8); and David saying, "I am a worm, less than human" (Psalms 22:7). On the other hand, non-Jews aggrandizing themselves include Pharoah, who says "Who is Adonai [the name of God] that I should listen to Him?" (Exodus 5:2); Sanherib, who says, "Which among all the gods of those countries [Hamath, Arpad, Sepharvaim, Hena, Ivvah] saved their countries from me, that Adonai should save Jerusalem from me?" (2 Kings 18:35); Nebuchadnezzar, whom Isaiah quotes as saying, "I will climb to the sky; higher than the stars of God I will set my throne. . . . I will match the Most High" (Isaiah 14:13–14); and Hiram, King of Tyre, who asserted, "I am a god; I sit enthroned like a god in the heart of the seas" (Ezekiel 28:2). This story, then, articulates one part of what must be built into a Jewish perspective of enhancing the human germline, namely, a significant degree of humility with regard to what we know, what we can do, and, most important, what we can know about the implications of what we can do.[33]

**God's Mold and the Uniqueness of Each Person**

Another important part of any Jewish stance on germline enhancement is a Mishnah (instruction of oral law) that has become justly famous. In describing the warning that the judges give to witnesses in a capital case to ensure that they give truthful testimony, the Mishnah says this:

> Therefore was a single man [Adam] created, to teach you that anyone who destroys a single person from the children of man is considered by Scripture as if he had destroyed an entire world, and that whoever sustains a single person from the children of man is considered by Scripture as if he sustained an entire world; and for the sake of peace among people, that no one could say to his fellow, my ancestor was greater than your ancestor . . . and to proclaim the greatness of the Holy One, blessed be He, for man stamps many coins with the same die and they are all alike, one with the other, but the King of kings of kings, the Holy One, blessed be He, stamps every person with the dies of the first person [in that every person is created in the divine image], and not one of them is like his fellow.[34]

In the last clause, the Mishnah graphically asserts the uniqueness of each and every person, and in the first clause, the worth of each and every person. The two of course are related, for just as an original Picasso is worth more than any of a hundred prints of it and much more than one of a thousand photographs of it, so too the uniqueness of each individual makes him or her all the more valuable than if there were many exactly like him or her. Therefore, if germline modification in any way undermines the uniqueness or worth of the people who are its products, it is to be denied theological and moral legitimacy on those grounds. If we are to engage in germline therapy, we must at a minimum take precautions to ensure that the people who are thereby produced are treated with a dignity equal to that of those born without such technology, recognizing them as the distinctive and infinitely worthwhile individuals they are.

**Rabbi Johanan and the Mikveh**

Nevertheless the Jewish tradition does have a few cases in which people tried to influence the genomic structure of animals or humans, and they were not condemned for this. As Jacob was preparing to depart from Laban after serving him for twenty years, he made a deal with Laban according to which Jacob would get the striped goats and dark-colored sheep and Laban the white or dark ones. Basing himself on the folklore belief of the time that goats and sheep seeing striped rods during mating would bear striped young, Jacob arranged for the sturdier animals to mate in the sight of such rods but not the weaker ones, thus producing for him both more and stronger animals.[35] The Torah does not denounce Jacob for his trickery; in fact in the next chapter, when Laban catches up with Jacob and complains that Jacob had stolen both his daughters and his flocks, Jacob asserts the righteousness of his actions.[36]

Later interpreters are concerned about Jacob's honesty; David Kimhi (=Radak, France, 1160?–1235?), for example, asserts in his commentary on this passage that Jacob did not resort to his method of ensuring that the animals would be speckled or striped the first year after the agreement with Laban, and thereafter he used his method only with his own flock, for otherwise he would have been engaging in flagrant dishonesty. More important for our purposes, however, is the fact that the Torah does not object to Jacob's intervention in the natural mating processes to produce animals of a specific color or body structure, and neither do medieval or modern Jewish commentators. Jacob in fact attributes his knowledge of how to change the usual color of the animals from white or dark to speckled to God Himself: "Had not the God of my father, the God of Abraham and the Fear of Isaac, been with me, you [Laban] would have sent me away empty handed. But God took notice of my plight and the toil of my hands, and He gave me judgment last night [when Jacob fled with his wives and speckled animals]."[37] One could thus argue either that God granted Jacob the knowledge and permission to alter the animals' color only to overcome an injustice but not as a matter of course; or alternatively one could assert that this story shows that in general God is not opposed to human beings causing such genetic changes. Thus no clear lesson can be drawn from this text.

Another talmudic story, however, is more directly on point in two ways: It deals with people who try to influence the genetic structure of human offspring, and the rationale is not justice but solely the creation of beautiful children. "Rabbi Yohanan was accustomed to go and sit at the gates of the *mikveh* [the specially constructed bathing place used by women to mark the completion of their menstrual cycle and thus the renewed permission to have sexual relations with their husbands]. He said: 'When the daughters of Israel come up from bathing, they will look at me and have children as handsome as I am' "[38] The Talmud objects neither to the unmitigated ego involved in what he did nor to his use of what he construes to be an effective technique to alter genetic structure; it is only worried that in seeing these women he will be tempted to have sex with them himself, and to that concern it gives several textual proofs that "the evil eye" has no power over him or any of the other descendants of Joseph. Thus it appears from this text that we may not only attempt to cure diseases through medical care, but we may even try to use genetic technology to enhance offspring—in this case, to make them more handsome.

If the Talmud knew about genes, would it have condoned more invasive therapies than sitting outside the ritual bath—techniques like genetic modification? If so, for what purposes would it approve of such measures other than beauty? Unfortunately, this text provides no guidance on these matters, but it does seem to provide a strong grounding to assert the legitimacy of genetic modification.

**The Use of Magic**

The Torah prohibits the use of magic,[39] which is understood as techniques intended to force God to do or make something, in contrast to prayers asking God to do so. The rabbis thus needed to define what constitutes violation of that ban and what does not, a task they carry out with regard to all of the Torah's prohibitions. What is the line distinguishing magic from permitted and even mandated human inventions intended to fix the world?

The Mishnah (edited in approximately 200 C.E.) says this: "If a sorcerer performs an act, he is subject to penalties, but not if he merely creates illusions."[40] The Mishnah is clearly not concerned with those who do magic tricks as a form of entertainment, for everyone watching understands that the magician is somehow using the laws of nature to produce an illusion, however much they are surprised by it and befuddled about how the magician pulled it off. What concerns the Mishnah instead are those who challenge the authority of God to control nature. The penalties for doing that are severe—violators are stoned to death!—and that indicates the seriousness with which both the Torah and rabbis approach this subject.

The Talmud (edited c. 500 C.E.), which is the later exposition of the Mishnah, gives us the details. One portion of the Talmud's discussion that is important for our purposes is this: "Some actions are entirely permissible, like the one of Rabbi Hanina and Rabbi Oshaya, who every Sabbath evening studied the doctrine of creation, by means of which they created a half-grown calf and ate it."[41] Why was this action permitted? In part, because it was done by rabbis studying the world that God had created, here through what they took to be God's description of that creation in the Torah. One is not forcing God's hand when one uses the very mechanisms that God inserted in what He made. This implication is further supported by another part of the Talmud's discussion: "It is different when it is in order to learn. 'You must not learn *to do* [abhorrent things]' (Deuteronomy 18:9). [This means that] you should not learn them in order to practice them, but you must learn to do everything in order to understand and to teach."[42]

Another important aspect of why the work of Rabbis Hanina and Oshaya was permissible is because they did it in consort: it was public and shared, not private and secret. That the Talmud intends this implication is demonstrated by the story it includes in this context about Rabbi Eliezer's death. Rabbi Eliezer was known for his expertise in what was pure and impure. Even so the rabbis, in another talmudic story, insist that the law is to be determined by the majority of them and not by a single scholar, expert though he is. Rabbi Eliezer gets a river to flow backward and a tree to move, but that does not convince his colleagues of his ruling; in fact, even a voice from Heaven declaring that Rabbi Eliezer is always right in such matters, a voice that the rabbis are sure is authentically God's voice, does not persuade them that the law should follow Rabbi Eliezer, for as the Torah itself says, "it [the law] is not in heaven"(Deuteronomy 30:12), a passage the rabbis daringly cite to quote God against Himself. Because Rabbi Eliezer did not submit to the authority of the rabbinic majority, he was excommunicated.[43]

Now, at his death, he is angry, defiant, and yelling at his son about the rules of the Sabbath, the one that will be his last. In this scene he also expresses his outrage for being abandoned by his students and mourns the loss of all the things he has discovered and cannot live to pass on: “Much Torah have I learned, and yet my disciples have only drawn from me as much as a painting stick from its tube.” Some of that knowledge, interestingly, is directly related to medicine: “Moreover, I have studied three hundred laws on the subject of a deep bright spot [which, according to Leviticus 12:2, is one of the signs of leprosy], and yet no man has ever asked me about them.” His knowledge also includes manipulating nature through magic: “Moreover, I have studied three hundred (or, as others state, three thousand) laws about the planting of cucumbers [by magic], and no man, except [Rabbi] Akiba ben Joseph, ever questioned me about that.”[44] Despite his students’ certainty that his expertise in the law about these matters greatly surpassed theirs, his insistence on determining the law by himself required that he not be listened to until ultimately he dies, when Rabbi Joshua lifts the sanctions against him and his teachings. Knowledge must be discussed and evaluated in public to be accepted, for otherwise even it is correct it can be dangerous and idolatrous.

Professor Laurie Zoloth, in a probing application of this talmudic passage, notes that the Jewish tradition places a strong emphasis on justice, requiring that any therapeutic products of genetic research and manipulation be available to everyone regardless of status or income. She then summarizes the lessons that she draws from this passage of the Talmud:

> We can develop a policy of what I call “civic witness,” allowing for the fullest freedom for basic research and the careful apprehension about any applied clinical experiments. In this, I argue for accountable, witnessed, and discussed research, for the public discourse and for the scholarly argument, beginning at the bench of basic science. Maximal freedom can be fully supported—“all knowledge in order to learn”—but fully learned in order to be fully taught and explained and held to the duty of justice. A witnessed and burdened freedom.
>
> . . . We stand in a world clamorous with need, and it is not only the need for genetics, it is the need for justice, health care access, basic public health, and nursing care. It is a world fraught with a troubling history of eugenics and with the eager pharmaceutical marketplace, and some are happy to privatize the research. On some level, faced with the reality of human suffering, pediatric disease, and devastating degenerative illness, gene therapies cannot come too quickly; and on some level, faced with our propensity for mistakes, our history

of eugenics, and our already unfair standing [in society generally and in health care in particular], it needs a far slower pace and the steady interlocutors of the respondent community. If it cannot answer us [about any specific procedure], we are alerted to trouble.[45]

## Maimonides' Prescription for Life

On the theological level, the body is God's creation as much as the mind, emotions, will, and spirit are. Like all our other faculties, our body should be used to live a life of holiness by obeying God's commandments. Thus Maimonides, a famous twelfth-century rabbi, physician, and philosopher, warns us about improper goals for life and describes what we should be striving to do in our lives:

> He who regulates his life in accordance with the laws of medicine with the sole motive of maintaining a sound and vigorous physique and begetting children to do his work and labor for his benefit is not following the right course. A man should aim to maintain physical health and vigor in order that his soul may be upright, in a condition to know God. . . . Whoever throughout his life follows this course will be continually serving God, even while engaged in business and even during cohabitation, because his purpose in all that he does will be to satisfy his needs so as to have a sound body with which to serve God. Even when he sleeps and seeks repose to calm his mind and rest his body so as not to fall sick and be incapacitated from serving God, his sleep is service to the Almighty.[46]

Whether this directive to engage in normal human activities to preserve health should drive us to seek germline cures for diseases that make it difficult for a person to serve God is, as one would guess, a complicated question. Certainly the strong Jewish commitment to medicine would argue that we should engage in genetic manipulation to cure or prevent diseases if we can. At the same time, we must take account of the first part of what Maimonides says—namely that the goal of life is not solely to have children and a health of body to function more effectively in achieving our human, utilitarian goals.

There is nothing wrong with trying to make human lives healthier, more just, and richer in meaning; on the contrary, that is one way of stating the Jewish mission to fix the world. In doing that, however, we must keep in mind that our ultimate purpose must be to serve God, and we do that at least as much when we tend to the sick and needy as when we develop cures for their ailments. Whether we are trying to change the world to improve it or whether we are acting in the world as it is to bring more caring, health, and justice, the critical thing to remember is that we must always act with what we construe to be God's purposes for our lives. For Jews, that is articulated in Jewish theology and law, Jewish stories and history, Jewish proverbs and prayers.

## "Choose Life"

We live in a time of great genetic promise and danger: promise that new research will prevent some human ailments altogether and cure others, and danger that in the process we will create physical, social, or moral monstrosities. Christians speaking about these prospects sometimes cite their doctrine of original sin, using it to warn us to be wary of the human penchant to use our new powers to sin. The Jewish tradition does not believe in original sin any more than it believes in original virtue. We are instead created morally neutral, with the ability to do both good and bad and knowledge of the difference. Because God is good and loving, he does not just throw us out into the world to fend for ourselves. He gives us a book of instruction, the literal meaning of the word "Torah," so that we can use our powers to choose good over bad, life over death. In these times, then, these ancient words in the Torah have special meaning: "I call heaven and earth to witness before you this day: I have put before you life and death, blessing and curse. Choose life—if you and your offspring would live—by loving the Lord your God, heeding His commands, and holding fast to Him."[47] May we be moral and wise enough to learn from our traditions how to protect ourselves from our selfish and destructive instincts and how to maximize instead our altruistic and constructive abilities in this new world of genetic challenges and hope.

## Notes

M.: Mishnah (edited by Rabbi Judah, president of the Sanhedrin, c. 200 C.E.)

J.: Jerusalem (or Palestinian or western) Talmud (edited c. 400 C.E.)

B.: Babylonian Talmud (edited by Ravina and Rav Ashi, c. 500 C.E.)

M. T.: Maimonides' *Mishneh Torah*, completed in 1177 C.E.)

S. A.: Joseph Karo's *Shulhan Arukh*, completed in 1563, with later glosses by Moses Isserles

1. For a description of some of the major ways in which these three lenses agree and disagree with each other, see my book, *To Do the Right and the Good: A Jewish Approach to Modern Social Ethics* (Philadelphia: Jewish Publication Society, 2002), ch. 1.

2. For a description of how each of these factors plays a role in making Jewish moral decisions and the advantages and disadvantages of using law for that purpose, see the appendix of my book, *Love Your Neighbor and Yourself: A Jewish Approach to Personal Ethics* (Philadelphia: Jewish Publication Society, 2003).

3. See, for example, James Drane, *More Humane Medicine: A Liberal Catholic Bioethics* (Edinboro, PA: Edinboro University Press, 2004). For the range of Catholic opinions on a variety of issues in bioethics, together with a discussion of a similar range among Jewish writers, see Aaron L. Mackler, *Introduction to Jewish and Catholic Bioethics: A Comparative Analysis* (Washington, DC: Georgetown University Press, 2003).

4. For some sample Jewish methodologies to address modern moral issues, see the articles by Borowitz, Israel, Ellenson, Newman, Dorff, Mackler, and Zoloth-Dorfman in *Contemporary Jewish Ethics and Morality: A Reader*, Elliot N. Dorff and Louis E. Newman, eds. (New York: Oxford University Press, 1995), chs. 7–12, 15.

5. In my book on Jewish medical ethics, I identify and discuss seven such underlying principles; see Elliot N. Dorff, *Matters of Life and Death: A Jewish Approach to Modern Medical Ethics* (Philadelphia: Jewish Publication Society, 1998), ch. 2.

6. Deuteronomy 10:14; Psalms 24:1.

7. That we use some of our property for worshiping God with sacrifices: e.g., Leviticus, chs. 1–5. That we use some of our property to help the poor: Leviticus 19:9–10; 25:33–38; Deuteronomy 15:7–8; and, especially, Leviticus 25:23: "But the land must not be sold beyond reclaim [by the person from whom it was taken to pay off a debt], for the land is Mine; you are but strangers resident with Me."

8. See, for example, Deuteronomy 10:14; Psalms 24:1.

9. Thus, for example, bathing is a commandment, according to Hillel: *Leviticus Rabbah* 34:3. Maimonides includes rules requiring proper care of the body in his code of Jewish law as a positive obligation (not just advice for feeling good or living a long life), parallel to the positive duty to aid the poor: M.T. *Laws of Ethics* (*De'ot*), chs. 3–5.

10. B. *Shabbat* 32a; B. *Bava Kamma* 15b, 80a, 91b; M.T. *Laws of Murder* 11:4–5; S.A. *Yoreh De'ah* 116:5 gloss; S.A. *Hoshen Mishpat* 427:8–10. Jewish law views endangering one's health as worse than violating a ritual prohibition: B. *Hullin* 10a; S.A. *Orah Hayyim* 173:2; S.A. *Yoreh De'ah* 116:5 gloss.

11. One could ask of course what the state could do to you after you committed suicide. The truth is, though, that the state could prevent people who commit suicide from passing on their estate to their heirs, a serious consequence. States could even publicly shame you and your family by denying you a proper burial, as Jewish law officially does, although in Jewish law every suicide is ruled as temporarily insane, therefore not responsible for his or her actions, and consequently is eligible to be buried according to the usual rites.

12. That suicide is legal in every state, but assisting a person to perform that legal act is illegal in all states except Oregon is, if I may say so, a rather curious development in American law. The Ninth and Second Circuit courts declared that assisting a suicide should be permitted as a constitutional right, based either on the liberty clause (the Ninth Circuit) or the equal protection clause (the Second Circuit) of the Fourteenth Amendment; see *Compassion in Dying v. State of Washington* 79. F.3d 790 (9th Cir. 1996) and *Quill v. Vacco* 80 F.3d 716 (2d Cir. 1996). I think that that articulates well the general American understanding of our rights over our bodies as declared by *Roe v. Wade* (1973, a woman's right to an abortion), *Nancy Cruzan* (1990, a person's right to refuse treatment), and other Supreme Court rulings. The U.S. Supreme Court, however, by a vote of 9-0, overruled those circuit court rulings, maintaining that this is not a matter covered by the Constitution but rather falls under the jurisdiction of state laws; *Washington v. Glucksberg* 117 S.Ct. 2258 (1997); *Quill v. Vacco* 117 S.Ct. 2293 (1997).

13. B. *Sanhedrin* 91a–91b. Also in *Mekhilta*, Beshalah, Shirah, ch. 2 (Saul Horowitz and I. A. Rabin, eds., 1960, p. 125); *Leviticus Rabbah* 4:5.

14. M. *Avot (Ethics of the Fathers)* 2:2.

15. For an eloquent articulation of this point, see Abraham Joshua Heschel, "The Patient as Person," in his book, *The Insecurity of Freedom: Essays on Human Existence* (Philadelphia: Jewish Publication Society, 1966), pp. 24–38.

16. B. *Shabbat* 54b. Along with Jeremiah (31:29–30) and Ezekiel (18:20–32), this offends our sense of justice, but that is only because we are so used to thinking in individualistic terms.

17. God causes illness as punishment for sin: Leviticus 26:14–16; Deuteronomy 28:22, 27, 58–61. God is our healer: Exodus 15:26; Deuteronomy 32:39; Isaiah 19:22; 57:18–19; Jeremiah 30:17; 33:6; Hosea 6:1; Psalms 103:2–3; 107:20; Job 5:18.

18. B. *Bava Kamma* 85a bases the permission to heal on the Torah's requirement that assailants provide for the recovery of their victims in Exodus 21:19. B. *Bava Kamma* 81b maintains that we not only have permission to heal, but the duty to do so based on the duty to return lost objects to their owners in Deuteronomy 22:2. B. *Sanhedrin* 73a uses Leviticus 19:16 as the ground for our duty to save lives and also to spend money to hire others to do so when we do not have the required expertise.

19. S. A. *Yoreh De'ah* 336:1.

20. J. *Kiddushin*; 66d; B. *Sanhedrin* 17b.

21. According to Immanuel Jakobovits, quoting the historian Cecil Roth, no less than half of the best-known rabbis in the Middle Ages, as well as poets and philosophers, were physicians by profession; see Immanuel Jakobovits, *Jewish Medical Ethics* (New York: Bloch, 1975), p. 205; Cecil Roth, *The Jewish Contribution to Civilisation* (Oxford: Oxford University Press, 1943), p. 192. For more on this, see, for example, Harry Friedenwald, *The Jews and Medicine* (Baltimore: Johns Hopkins University Press, 1944; reprinted by New York: Ktav, 1967), 2 vols.; Michael Nevins, *The Jewish Doctor: A Narrative History* (Northvale, NJ: Jason Aronson, 1996).

22. *Midrash Temurrah* as cited in *Otzar Midrashim*, J. D. Eisenstein, ed. (New York, 1915) II, 580–581. See also B. *Avodah Zarah* 40b, a story in which a rabbi expresses appreciation for foods that can cure. Although circumcision is not justified in the Jewish tradition in medical terms, it is instructive that the rabbis maintained, as noted earlier, that Jewish boys were not born circumcised specifically because God created the world so that it would need human fixing, an idea similar to the one articulated here on behalf of physicians' activity despite God's rule; see *Genesis Rabbah* 11:6; *Pesikta Rabbati* 22:4.

23. B. *Sanhedrin* 73a. Note that in contrast, in American law only Wisconsin and Vermont make helping others in distress a positive duty, with failure to do so a misdemeanor punishable by a fine of not more than $100; in fact, until many states recently passed "Good Samaritan laws," a person who unintentionally harmed someone in the attempt to save him or her could actually be sued.

24. B. *Sanhedrin* 84b (on the permission to inflict pain in order to heal), 73a (on the requirement to spend money to heal when we lack the expertise); Nahmanides, *Torat Ha-Adam, Sha'ar Sakkanah*, quoted by Joseph Karo, *Bet Yosef*, *Yoreh De'ah* 336.

25. While the tradition applies these verses only to fellow Jews, it nevertheless requires that Jews supply health care to non-Jews as well for the sake of good relations, even though historically non-Jews, far from providing health care for Jews, often persecuted, maimed, and killed them. That context makes this provision of Jewish law nothing less than remarkable. See B. *Gittin* 61a.

26. M. *Avot* (*Ethics of the Fathers*) 1:2.

27. *Los Angeles Times*, April 13, 1988, pp. A1, 14, 15. As reported in the *Los Angeles Times*, February 1, 2003, Part 2, p. 23, a later poll conducted by the American Jewish Committee in 2003 asked 1,008 Jews to choose the quality most important to their Jewish identity; 41 percent said "being part of the Jewish people," 21 percent said "commitment to social justice," and only 13 percent chose "religious observance."

28. Psalms 115:16.

29. *Ecclesiastes [Kohelet] Rabbah* 7:19; see also *Midrash Zutah, Ecclesiastes [Kohelet]* 7:11. For a thorough discussion of Jewish sources on preserving the environment, see Arthur Waskow, ed., *Torah of the Earth: Exploring 4,000 Years of Ecology in Jewish Thought* (Woodstock, VT: Jewish Lights Publishing, 2000), 2 vols.; and Ellen Bernstein, ed., *Ecology and the Jewish Spirit* (Woodstock, VT: Jewish Lights Publishing, 1998).

30. *Genesis Rabbah* 11:6; *Pesikta Rabbati* 22:4.

31. Genesis 11:7.

32. B. *Hullin* 89a.

33. For a nice discussion of the value of humility in Jewish thought and law, see Sol Roth, "Toward a definition of humility," *Tradition* 14 (1973–1974): 5–22; reprinted in Dorff and Newman, eds., *Contemporary Jewish Ethics and Morality*, pp. 259–270.

34. M. *Sanhedrin* 4:5.

35. Genesis 30:37–43.

36. Genesis 31:36ff.

37. Genesis 31:42.

38. B. *Berakhot* 20a.

39. For example, Leviticus 20:27; Numbers 23:23; Deuteronomy 18:9–22; 1 Samuel 15:23; 2 Kings 21:6; 2 Chronicles 33:6.

40. M. *Sanhedrin* 7:11 (67a).

41. B. *Sanhedrin* 67b.

42. B. *Sanhedrin* 68a.

43. B. *Bava Metzia* 59b.

44. B. *Sanhedrin* 68a.

45. Laurie Zoloth, "Reasonable magic and the nature of alchemy: Jewish reflections on human embryonic stem cell research," *Kennedy Institute of Ethics Journal* 12:1 (March 2002): 65–93; given originally as the Franck Lecture in Bioethics at Georgetown University. The quotations appear on pp. 92–93.

46. M. T. *Laws of Ethics* (*Hilkhot De'ot*) 3:3.

47. Deuteronomy 30:19.

# 3

# The Roman Catholic Magisterium and Genetic Research: An Overview and Evaluation

Thomas A. Shannon

Genetic engineering and research have been at the center of attention for several decades. Ever since the discovery of the DNA molecule by Watson and Crick in 1953 there has been relatively steady progress both in understanding the composition of the human genetic code and intervening in it. With the completion of the Human Genome Project, scientists are coming closer to their goal of developing genetic therapies for many of the diseases that continue to plague humanity.

However, along with this scientific progress there are significant ethical questions about the research itself, the immediate consequences for human subjects, and how such research fits into larger questions of health care policy and social justice. That this question is important is evidenced by the multitude of international and national norms regulating research on humans. Specifically in the United States, there have been presidential commissions to study this topic and to propose regulations. There is an elaborate network of institutional review boards to evaluate research on human subjects. There is an officer of research regulation in the U.S. Department of Health and Human Services. And in light of recent deaths of volunteers in research protocols, attention is being focused on both the adequacy of the norms as well as their enforcement.

Given the importance of this topic, it should come as no surprise that the Roman Catholic Church would have a body of teaching on this topic. Medical ethics has been an important part of the moral tradition from at least the Middle Ages forward, and in contemporary times this tradition has been developed in response to new questions, including research on human subjects. In this chapter I present the view of the Roman Catholic highest authority, or Magisterium, on genetic research to identify the ethical issues that it considers important. Second, I evaluate that position and in conclusion, suggest some other perspectives on the problems and solutions raised by the Magisterium.

## The Basis of the Position

In this section I identify the general framework for research in ethics. The focus here is setting up the general context in which the teaching is developed and the general principles that serve as the framework.

*Donum vitae*, for example, identifies the "criteria of moral judgment as regards the applications of scientific research, especially in relation to human life and its beginnings. These criteria are the respect, defense and protection of man, his 'primary and fundamental right' to life, his dignity as a person who is endowed with a spiritual soul and with moral responsibility and who is called to beatific communion with God."[1] This perspective is augmented by the *Catechism of the Catholic Church* in stating that "science and technology by their very nature require unconditional respect for fundamental moral criteria. They must be at the service of the human person, of his inalienable rights, of his true and integral good, in conformity with the plan and will of God."[2]

These general criteria give rise to the observation that "Thus science and technology require for their own intrinsic meaning an unconditional respect for the fundamental criteria of the moral law: That is to say, they must be at the service of the human person, of his inalienable rights and his true and integral good according to the design and will of God."[3] Following up on this theme, the Pontifical Academy for Life notes that "The most urgent need now seems to be that of re-establishing the harmony between the demands of scientific research and indispensable human values. The dignity of scientific research consists in the fact that it is one of the richest resources for humanity's welfare."[4] These comments highlight the positive esteem in which science and research are held by the church. These perspectives open up new areas of understanding and provide new opportunities of being of service to men and women of the entire world, and such opportunities are valued greatly.

The *Catechism of the Catholic Church* notes that "Basic scientific research, as well as applied research, is a significant expression of man's dominion over creation. Science and technology are precious resources when placed as the service of man and promote his integral development for the benefit of all. . . . Science and technology are ordered to man, from whom they take their origin and development; hence they find in the person and in his moral values both evidence of their purpose and awareness of their limits."[5]

Thus the core of the ethics of research and the uses to which science and technology are to be put are rooted in the very nature of the person and of the created order. Participation in the project of science and research help achieve our final fulfillment, as noted by the Pontifical Academy for Life: "In summary, therefore, there should be a reaffirmation of the right and duty of man, according to the mandate from his Creator and never against the natural order established by him, to act within the created order and on the created order, making use as well, of other creatures, in order to achieve the final goal of all creation: the glory of God and the full and definitive bringing about of this Kingdom, through the promotion of man."[6]

## The Human in Research

The focal point of this chapter is research on human subjects. One key element in the research ethic proposed by the Magisterium is the centrality of the human in the development of the ethic. Thus in *Donum vitae* we read: "An intervention on the human body affects not only the tissues, the organs and their functions, but also involves the person himself on different levels. . . . Thus, in the body and through the body, one touches the person himself in his concrete reality."[7] This is a critical observation because it is a rejection of a Platonic or Cartesian dualism that could justify a variety of interventions because they are done on the body, not on the person or on the self. The statement highlights the unity of the person and the necessity of realizing that touching the body initiates an engagement with the person, an engagement that is of necessity couched in an ethical perspective. While it is the case—and indeed often a necessity—that we consider the objectivity of the body, as we do with various imaging technologies and in surgery, nonetheless if we allow this objective perspective to become the dominant or only perspective, then we will have marginalized the core ethical reality of the human and the foundation of a research ethic.

Pope John Paul II notes the danger of forgetting this intimate corporeal subjectivity that is characteristic of the human: "*From being a subject and goal*, man is not infrequently considered an object and even a form of 'raw material'; here we need only mention experiments in genetic engineering which are a source of great hope but at the same time of considerable preoccupation for the future of the human race."[8] Of significance in the Pope's perspective is the clear value of the research or of the particular intervention; this is never denied. What is singled out though is the necessity of balance in one's perspective. The valued drive for knowledge and understanding achieved through research needs to be put into a context in which that drive is both nurtured and shaped by an ethical perspective.

For the Magisterium, the human reality begins with conception: "From the moment of conception, the life of every human being is to be respected in an absolute way because man is the only creature on earth that God has 'wished for himself' . . . No one can in any circumstance claim for himself the right to destroy directly an innocent human being."[9] The *Catechism of the Catholic Church* articulates this position: "Human life must be respected and protected absolutely from the moment of conception. From the first moment of his existence, a human being must be recognized as having the rights of a person—among which is the inviolable right of every innocent being to life."[10]

This position clearly has significant consequences for our later discussion of specific forms of genetic research, but it is a position coherent with understanding the human as the center of the ethical universe. This position coheres with the understanding of the human as an integrated whole. Although one is manipulating cells, one is still, from this perspective, touching the human. In speaking of the Universal Declaration of Human Rights, the Vatican observes, "Thus it does not define the bearers of the rights which it proclaims; it does not affirm that these rights belong to every human being from the moment when he or she emerges as an individual from his or her very genetic heritage. . . . The fact that unborn human beings and human embryos are not explicitly protected opens the door, particularly in the field of genetic intervention, to the very forms of discrimination and the violations of human dignity which the Declaration seeks to ban."[11] Clearly this perspective curtails some forms of research, but the perspective is grounded in a coherent vision of the human person and the subjectivity of the human body.

**Permissible Research**

The Magisterium deals with the issue of research in genetics by focusing on particular issues. The general principles presented here serve as a general background for the discussion and other principles or perspectives are brought into the discussion as needed.

*Donum vitae* addresses the morality of prenatal diagnosis by saying "*If prenatal diagnosis respects the life and integrity of the embryo and the human fetus and is directed toward its safeguarding or healing as an individual, then the answer is affirmative.*"[12] Further, the teaching states "Such diagnosis is permissible, with the consent of the parents after they have been adequately informed, if the methods employed safeguard the life and integrity of the embryo and the mother, without subjecting them to disproportionate risks."[13] In addition *Donum vitae* argues that "Given that it is a question of research, and therefore a very restricted intervention on the patient it can be acceptable, provided that 'it is not otherwise possible' and, if the subject is unable to give consent, that further conditions are met: minimal risk, consent by those whose legal right it is to give it, undoubted advantage for the health of persons in the same category, lack of other resources and possibilities for research."[14] The *Catechism of the Catholic Church* compliments this perspective by arguing that "One must hold as licit procedures carried out on the human embryo which respect the life and integrity of the embryo and do not involve disproportionate risks for it, but are directed toward its healing, the improvement of its condition or health, or its individual survival."[15]

In general one can see these three positions as containing core elements of the mainstream discussions of research ethics: consent of those responsible for the child, avoidance of disproportionate risks for both fetus and mother, a therapeutic intent, and the hope of benefit for others. Again, in general, one can argue that the citation from the *Catechism* in the preceding paragraph is a relatively mainstream ethical framework for fetal research.

This is the general case. The position of the Catholic Church, as indicated here, affirms the presence of a person from conception forward. This is what gives the Magisterium's position on embryonic or fetal research a particular shape and differentiates it from the position of many other Christian religions, some federal guidelines, the research guidelines of other countries such as the United Kingdom, and some narrowly drawn cases of federal funding for embryonic stem cell research.

## General Norms

According to *Donum vitae*, "*Medical research must refrain from operations on live embryos, unless there is a moral certainty of not causing harm to the life or integrity of the unborn child and the mother, and on condition that the parents have given their free and informed consent to the procedure.*"[16] The document continues: "In the case of experimentation that is clearly therapeutic, namely, when it is a matter of experimental form of therapy used for the benefit of the embryo itself in a final attempt to save its life and in the absence of other reliable forms of therapy, recourse to drugs or procedures not yet full tested can be licit."[17] Here again research is morally justified as long as it is part of a therapeutic intervention, perhaps a justification of using a not-yet-validated experimental intervention on the basis of compassion. Again this is well-accepted ethical research practice.

However there can be a danger of stretching the compassionate use criterion too far. In addition the fetus might be seen as an ideal research candidate and its interests might be placed second to the research agenda. Here the *Catechism* notes that "It is an illusion to claim moral neutrality in scientific research and its applications. On the other hand, guiding principles cannot be inferred from simple technical efficiency, or from the usefulness accruing to some at the expense of others or, even worse, from prevailing ideologies."[18] At this stage this is an appropriate warning not to make the fetus into a means to an end or to subsume its well-being under the vague promise of benefit to future generations.

To shore up this position morally, the *Catechism* reminds us of the core ethical principles that have been articulated: respect for the dignity of the person (for the Catholic Church, this includes embryos and fetuses) and consent. "Research or experimentation on the human being cannot legitimate acts that are in themselves contrary to the dignity of persons and to the moral law. The subjects' potential consent does not justify such acts. Experimentation on human beings is not morally legitimate if it exposes the subject's life or physical and psychological integrity to disproportionate or avoidable risks. Experimentation on human beings does not conform to the dignity of the person if it takes place without the informed consent of the subject or those who legitimately speak for him."[19]

## Genetic Interventions

The Pontifical Academy for Life observes that "The positive value of an understanding of the genome of the species, and also in some cases of that of the individual, must be recognized. However, no one has an absolute right to such knowledge. The positive value of the acquisition of general information is based not only on the value of scientific knowledge as such, but most of all on the possibility of the service it can render to the good of the person."[20] As John Paul II has stated: "A strictly therapeutic intervention whose explicit objective is the healing of various maladies such as those stemming from chromosomal defects will, in principle, be considered desirable, provided it is directed to the true promotion of the personal well-being of the individual."[21]

With these statements, the Pope and the Pontifical Academy for Life focus on one of the core areas of research in contemporary biology: genetics. The importance and value of the knowledge of human genetics is clearly acknowledged. What is interesting is that the value of the knowledge is explicitly tied to the good of the person. That is, genetic knowledge is not seen abstractly; rather, it is put into a framework of service and benefit to humanity. Perhaps that is why there is an additional claim that there is no absolute right to such knowledge. While for some this may be a rejection of the value of knowledge for its own sake, nonetheless this is also a recognition that genetic knowledge in particular is typically sought for immediate clinical application rather than being pursued for its own sake. In fact this observation might serve as an important corrective to the tendency in modern biology to pursue primarily applied research rather than basic research. Yet this is not an affirmation of a limit to knowledge, but rather a cautionary reminder that applications have consequences and some of these may be problematic. "In principle, therefore, there are no ethical limits to the knowledge of the truth, that is, there are no 'barriers' beyond which the human person is forbidden to apply his cognitive energy . . . but, on the other hand, precise ethical limits are set out for the manner the human being in search of the truth should act, since '*what is technically possible is not for that very reason morally admissible*'."[22]

## Specific Perspectives

To help specify some of the ethical concerns for research in genetics in general, an analogy with xenotransplantation is proposed. While xenotransplantation is not yet being done, research is continuing and many think that it is an appropriate solution to the shortage of organs for human transplantation. The Pontifical Academy for Life has considered the ethics of such transplantation and come to two conclusions. One is that the ethical limit of for xenotransplantation is "in the degree of change that it may entail in the identity of the person who receives it." Thus if the xenotransplant would significantly change the person's identity, the transplant would be prohibited. Second, "those organs which are seen as being purely *functional* and those with greater personalized significance must be assessed, case by case, specifically in relation to the symbolic meaning which they take on for each individual person."[23] Thus the critical ethical test is the symbolic meaning of the transplant for the person, with the implied suggestion that if the organ to be transplanted from an animal is essentially functional—kidney, heart, lungs, for example—there is no ethical problem. However, should there be xenotransplants that come closer to the person—brain tissue, for example—this might constitute some ethical problem if it could affect the person's identity. Although this transplant will result in a transgenic human, the procedure is not per se prohibited.

This example can be used as a way to think through some of the ethical issues in genetic research. Is the research essentially functional in that it repairs an organ or that it provides some compensation for the problems with an organ? If so, the genetic research would be judged morally acceptable. It is essentially like any other therapeutic intervention that seeks a cure or compensation for a diseased organ or system. On the other hand, the closer the research would come to developing an intervention that would affect someone's identity or change the self to a substantive degree, that research would be judged immoral. Again the dignity of the person is a key value here as well as the integrity of that person. Even here, however, forms of genetic research involving the human brain are not per se off limits because of the correlation of the self and brain. The issue remains one of functionality. If the research has the consequence of significantly intervening with a person's self, then that research would be prima facie off limits.

A second category used to help develop a perspective on genetic research is that of risk, which is considered by the Pontifical Academy for Life: "Risk—understood as an unwanted or damaging future event, the actual occurrence of which is not certain but possible—is defined by means of two characteristics: the level of probability and the extent of damage. . . . The extent of damage in contrast, is measured by the effects that the event produces. . . . Together, these two criteria—probability and extent of damage—define the *acceptability* of the risk, as reflected by the risk/benefit ratio. Only when a risk can be concretely assessed it is possible to apply criteria for evaluating its *acceptability*."[24] This is followed with advice on how to proceed when there is risk.

It is important to note that an absolute limit is not placed on research because of the presence of risk. "In the absence of data that allow a reliable assessment of such a risk, greater caution should be used; this does not necessarily mean, however, that a total 'block' should be put on all experimentation. . . . In this situation, therefore, the imperative ethical requirement is to proceed by 'small steps' in the acquisition of new knowledge, making use in experiments of the least possible number of subjects, with careful and constant monitoring and a readiness at every moment to revise the design of the experiment on the basis of new data emerging."[25] This position is harmonious with the research ethic in place in the vast majority of laboratories in the world. It is an ethical position of prudence, of carefully monitoring the effects of the research, and a willingness to redesign in light of new findings, expressed fittingly in the words of John Paul II: "*The Church respects and supports scientific research when it has a genuinely human orientation, avoiding any form of instrumentalization or destruction of the human being and keeping itself free from the slavery of political and economic interests.*"[26]

**Prohibited Research**

To this point, the overall argument of the Catholic Church has been in support of the research project in biomedicine and genetics. In many ways, with the exception of some of the expressly religious warrants appealed to, the research ethic proposed is harmonious with that held by many scientists and researchers: consent, protection of the subject, risks proportionate to the benefits expected to be gained. Nevertheless there is a difference because the Catholic Church draws a clear and firm line with some forms of research. The next section presents some general guidelines and then examines the cases of cloning and embryonic stem cell research.

**General Norms**

There are three statements that lay out quite clearly a moral framework of prohibited areas of research, derived from the positions previously described.

First, "*If the embryos are living, whether viable or not, they must be respected just like any other human person; experimentation on embryos which is not directly therapeutic is illicit.*"[27] The conclusion drawn is that "It is immoral to produce human embryos intended for exploitation as disposable biological material."[28]

Second, "The human body is an integral part of every individual's dignity and it is not permissible to use women as a source of ova for conducting cloning experiments."[29]

And third, "*Certain attempts to influence chromosomic or genetic inheritance are not therapeutic but are aimed at producing human beings selected according to sex or other predetermined qualities. Such manipulations are contrary to the personal dignity of the human being and his or her integrity and identity' which are unique and unrepeatable.*"[30]

These are limits on research that follow from the general principles, which include respect for human persons, not objectifying humans, and not exploiting them. The dividing line between the church's official position and that of the research ethics of the broader society has to do with the evaluation of the human embryo, as will become evident from the next two sections that describe research efforts using human embryos.

## Cloning

The cloning of the sheep Dolly was a biological breakthrough in that she was the first mammal to be cloned. And as is usual with major breakthroughs in biotechnology, there followed a prolonged debate in various media on the science and ethics of cloning. The lines of the debate are essentially twofold: The majority of scientists and researchers join with others in a desire to prohibit human cloning for reproductive purposes. This position is quite widespread across cultures, religions, and governments. The second debate focuses on cloning human embryos so that stem cells can be extracted and used in developing a therapy for the individual from whom the original cell was taken. This is usually called therapeutic cloning, but a better although a bit more clumsy term is cloning for therapeutic purposes. The cloning is not therapeutic, but is it done for a therapeutic purpose. Positions on this vary from absolute prohibitions to the British government's allowing this to proceed. The goal of such cloning is not reproductive but to clone an embryo from an individual and develop an appropriate cell line so that whatever cells or organs are implanted into the patient will not be rejected. This is truly individualized therapy.

The Magisterium rejects both reproductive cloning and cloning for therapeutic purposes. "A prohibition of cloning which would be limited to preventing the birth of a cloned child, but which would still permit the cloning of an embryo-foetus, would involve experimentation on embryos and foetuses and would require their suppression before birth—a cruel, exploitative way of treating human beings. . . . In any case such experimentation is immoral because it involves the arbitrary use of the human body (by now decidedly regarded as a machine composed of part) as a mere research tool. . . . It is immoral because even in the case of a clone, we are in the presence of a 'man,' although it is in the embryonic stage."[31]

With respect to research on embryos, which will be necessary for cloning for either reproductive or therapeutic purposes, the church says: "If the law allows unlimited human cloning for research, it will set the stage for further uses of the technique in humans and make a ban on cloning for live birth all but unenforceable here and now. . . . If human cloning is to be banned effectively, the ban must apply at the very beginning of the process."[32]

In addition, in an intervention delivered to the United Nations, the church argued against cloning on the basis of its consequences. "Cloning a human embryo, while intentionally planning its demise, would institutionalize the deliberate, systematic destruction of nascent human life in the name of unknown 'good' of potential therapy or scientific discovery. This prospect is repugnant to most people including those who properly advocate for advancement in science and medicine."[33]

## Embryonic Stem Cell Research

Embryonic stem cell research, which may or may not involve obtaining cells from cloned embryos, is also explicitly prohibited. Thus regardless of the origin of the embryos—post-IVF donation, cloning, or generated explicitly for research use—the Pontifical Academy for Life develops three reasons for the prohibition of such research: First, because a human subject is formed from the moment of the union of gametes, it is from then a human individual with its own right to life and "Therefore, the ablation of the inner cell mass (ICM) of the blastocyst, which critically and irremediably damages the human embryo, curtailing its development, is a *gravely immoral a*ct and consequently is *gravely illicit*."[34] Second, therapeutic cloning in conjunction with embryonic stem cell research is prohibited because human embryos are created in order to be destroyed to obtain the stem cells. Finally, this statement also prohibits using embryonic or differentiated cells derived from them because this "entails a proximate material cooperation in the production and manipulation of human embryos on the part of those producing or supplying them."[35]

The academy also notes that "the core of the debate on the protection of the human embryo does not involve identifying earlier or later indices of 'humanity' which appear after insemination, but consists rather in the recognition of fundamental human rights by virtue of the presence of a human being."[36]

Thus any form of research on the embryo that involves its destruction, its being used to benefit others, or its generation through cloning is prohibited. The reasoning for this prohibition follows directly from the church's position that the embryo is to be treated as a human person from the moment of its conception. Since one assumes a personal presence with the embryo, it then has the dignity of a person and must be protected as such.

## Genetic Modification of the Germline

Of all the controversial dimensions surrounding the debate over gene therapy, few are more problematic than germline gene therapy. The purpose of this therapy is to avoid a genetic disease in both the individual and in that individual's descendants. The prevention of the disease is the goal. The therapy is initiated at the embryonic level and prevents the disease from manifesting itself in the immediate patient as well as any future generations. Thus the benefits are both individual and familial or social. The risks, however, are significant. For the individual, if the therapy does not work, not only is there the possibility of being at risk for the disease, but also of other harms because of the failed genetic intervention. In addition, the change or changes that occur because of the intervention—whether or not they are successful—will be passed on to future generations if that individual reproduces. Thus the benefits are significant, as are indeed the risks.

On the question of human germline modification, a document issued by the Catholic Bishops of England in 1996 is of special interest. They argue that the genome is not untouchable because the "genome is simply one highly influential part of our bodies."[37] They conclude that like other parts of the body, "the genome may *in principle* be altered, to cure some defect of the body." Further, they argue that they could "imagine situations in which to choose this kind of treatment would be, not simply a *right* of the person choosing, but morally required."[38] The bishops conclude with a comment about the justification of germline research.

This position is also complemented by this statement of the International Theological Commission, "Communion and Stewardship: Human Persons Created in the Image of God." "Germ line genetic engineering with a therapeutic goal in man would in itself be acceptable were it not for the fact that is it is hard to imagine how this could be achieved without disproportionate risks especially in the first experimental stage, such as the huge loss of embryos and the incidence of mishaps, and without the use of reproductive techniques. A possible alternative would be the use of gene therapy in the stem cells that produce a man's sperm, whereby he can beget healthy offspring with his own seed by means of the conjugal act."[39]

While this is not a ringing endorsement of germline therapy—"it is hard to imagine how this could be achieved"—the objections are technical and ethical and in principle could be overcome. Should they be overcome, the research could in principle go forward. Granted that people should not be deprived without good reason of the genes they would otherwise have inherited from their parents and passed on to their children, the real possibility of eliminating from a family some serious disease—for example, Huntington's chorea—would appear to be good enough reason to improve on a person's genetic makeup and reproductive potential.[40]

These statements clearly must not be understood as carte blanche approval for research on human germline modification. The bishops note that the traditional ethical issues of consent, risks, and costs are relevant to this form of research. The major problem for them is that even though sometimes relatively high risks can be accepted, at the present time they judge the risks, particularly to the embryo, associated with germline gene therapy to go beyond what is reasonable. They also note that given the current state of knowledge and application, the therapy may cause undesired hereditary side effects, which might be difficult to undo and should therefore be prohibited at the present time.[41] Thus, to repeat the bishops, this is an in-principle argument. Whether such therapy will ever meet ethical standards is a separate question and one to be answered in light of future research. One could envision a case in which germline gene therapy is validated for a disease a fetus has. One could in principle argue that such therapy would be justified even though there might be germline modification. The intervention is therapeutic and the research is established. Some risks can be accepted as falling within the standard risks of interventions and the parents might accept these on behalf of their fetus.

## Evaluation and Conclusion

Two important elements emerge from the Catholic Church's general ethical position on research in genetics. First, the ethical foundation of permissible research is—absent a few explicit religious warrants—identical with mainstream research ethics: informed consent, an acceptable risk-benefit ratio, appropriate research design, the promise of benefit.

Second, the Magisterium is quite open to genetic research, even on embryos and fetuses, as long as the research is directed to a therapeutic end. Parents can consent to research on behalf of their embryo or fetus just as they can for an older child. The ethical touchstone is that this must be directed to some direct benefit for the subject. While undoubtedly some benefit for others will be derived from the research, the ethical basis of the research must be to benefit the subject. As the Pontifical Academy for Life notes: "Therefore any experimentation on the human embryo that does not have the goal of obtaining direct benefits for his/her own health, cannot be considered morally licit."[42] While this may be more narrowly drawn than the common research ethic, it does not stand totally outside the mainstream of research ethics.

This stance toward embryo research by the church is what divides it from the majority of the community, although by no means should the Catholic Church be thought of as the only group that opposes the use of embryos in research. The position of the church, which was described in more detail earlier, is that the embryo is to be treated as a person from conception. The church recognizes that there is no definitive philosophical argument that personhood begins at fertilization; it argues that since fertilization is the beginning of the process leading to personhood, the embryo must be treated as a person from the beginning. The embryo will have the dignity associated with a person and will have the rights associated with personhood. Thus Pope John Paul II draws the clear conclusion. "I condemn in the most explicit and formal way, experiential manipulation of the human embryo, since the human being, from conception to death, cannot be exploited for any purpose whatsoever."[43]

In evaluating this stance, I continue to argue that this perspective on personhood can be criticized. First, the position assumes that fertilization is a discrete moment, whereas it is process (as is the entire gestational sequence).[44] One cannot determine in advance a precise moment of fertilization. Rather, one can determine that the embryo is now in a different developmental stage. Second, the position identifies the establishment of the embryo's genetic code or genome as coincident with personhood. In addition to coming close to a form of genetic reductionism, the better description of the establishment of the embryo's genetic code is the establishment of the next genetic generation. Because the embryo is still in a developmental process, twinning may occur, perhaps triplets, and perhaps a twinned embryo may recombine. The critical variable here is that true biological individuality has not yet occurred. This is a function of the process of differentiation in which the cells of the developing embryo become committed to becoming specific body parts. This is true biological individuality and is certainly morally relevant with respect to thinking about philosophical individuality. My argument is that before one can be a person, one must first be an individual. Individuality is at least a biological, if not a philosophical, presupposition of personhood. Thus I conclude that the very early embryo, prior to the occurrence of differentiation, is better described as human nature.[45]

Thus I also conclude that within the larger process of embryogenesis, until the process of individuation is completed what we have is a living organism that has a certain biological and teleological unity and is genetically unique. What we have in this organism is human nature, nothing less but also, importantly, nothing more. This manifestation of human nature is unique, but it is not yet an individualized manifestation and a fortiori it is not yet a personal manifestation of that nature. There is a value to this organism because it is living, because it is in the process of development, because it is a unique manifestation of a life form, and because it has a degree of complexity and directionality. There is a functional unity in that this nature is the precursor of even more complexity and the ground of new modes of interaction within the larger environment, and because its cells are yet totipotent or pluripotent. There is a theological value in that this nature is within the created realm and is in the initial stages of a journey that will be completed at the conclusion of its life span.

However, these are not the values associated with individuals or persons. Clearly they are part of the values we associate with persons, but they are not the most important values associated with personhood. To return to the "image of God" discussion, the qualities of the person associated with this image are intellect and freedom. If anything, natures are not free; they act out their genetic plan. They fulfill their natures. However, persons can do more in that they can transcend their natures. In part they do this through the use of their intellect, but more significantly they do this through the use of their freedom and they do this more specifically in an act of love or commitment to the good. Their natures are transcended and transformed through this commitment to the value beyond themselves.

Persons, I argue, have a dignity; natures have a value. The dignity of the person grounds a more absolute standing, particularly with respect to interventions. The value of human nature does not generate the same level of protection as a person precisely because it is a nature. Nonetheless, it is human nature and it is to be valued.

This conclusion opens the possibility of research on human embryos such as cloning for therapeutic purposes or using embryos as the source for embryonic stem cell research. The destruction of embryos is indeed a significant disvalue because they are killed. Nonetheless because they are prepersonal, such killing is not murder. However, such a premoral disvalue needs to be offset by equally significant premoral and moral values. One problem with embryonic stem cell research is the context of exaggerated hyperbole in which it exists. If one listens to some of the rhetoric, one would think that the research is well advanced and clinical applications will start momentarily. Such is not the case. In addition although it is more difficult, complex, and expensive, there is the possibility of using adult stem cells as the basis of the research.

The ethical criteria for the destruction of embryos to obtain stem cells for research are (1) rigorous scientific protocols that are peer reviewed for merit (as is the case with other research), (2) extremely careful monitoring of the outcomes of the research (this would include dissemination of results and efforts at their replication), and (3) careful and limited clinical application to determine if in fact the expected benefits materialize. I think, therefore, a case can be made for a limited number of embryonic stem cell protocols. We need to determine if in fact this is a viable route. If specific cell lines cannot be developed as expected, this needs to be determined quickly so new directions can be determined. I also think such research should be funded by the federal government so that the protocols will be reviewed scientifically and ethically, as other research currently is.

Finally, there is an issue not addressed by the Catholic Church (nor many others, for that matter) with respect to research in genetics. How does research in genetics fit into an overall vision of health care in the United States? This is a question of priorities as well as resource allocation. Clearly there are many competing health care needs in the United States, but we have a great deal of trouble trying to establish any sense of priorities, particularly with respect to long-term needs. For example, where does preventive medicine fit into funding plans? While prevention will not eliminate disease, nonetheless a person who is healthier is a better patient than one who is not. Where do high-tech transitional technologies fit into the health care needs of the country? Where does research on genetic diseases fit in? Generally speaking, there is a lobby for almost every disease or program, and funding tends to be based on the effectiveness of the lobby rather than on the long- and short-term health care needs of the country. My position is that before vast research sums are expended on embryonic stem cell therapy, at least let there be a debate about both the scientific merits of the research and where such research might fit into the country's short- and long-term health care needs. Thus in addition to the important debate about the ethical status of the human embryo, there also needs to be a debate about the social justice implications of funding this research.

A comment from *Donum vitae,* which quotes from Vatican II, expresses very well a critical perspective that applies to how we think through different perspectives on research in general and research in genetics. "Science without conscience can only lead to man's ruin. 'Our era needs such wisdom more than bygone ages if the discoveries made by man are to be further humanized. For the future of the world stands in peril unless wiser people are forthcoming'."[46]

## Notes

1. The Holy See, *Donum vitae. Instruction from the Congregation of the Doctrine of the Faith*, Vatican, February 1987, I, 1.

2. United States Catholic Conference, *Catechism of the Catholic Church* (Washington, DC: United States Catholic Conference, 1994), p. 552.

3. *Catechism of the Catholic Church*, p. 552.

4. Pontifical Academy for Life, "Reflections on Cloning," Vatican, 1997, p. 6.

5. *Catechism of the Catholic Church*, p. 552.

6. Pontifical Academy for Life, "Prospects for Xenotransplantation: Scientific Aspects and Ethical Considerations," Vatican, 2001, p. 7.

7. *Donum vitae. Instruction*, I, 3.

8. John Paul II, "Address at the meeting with Rectors of Polish Universities," June 8, 1997. Italics in original.

9. *Donum vitae. Instruction*, I, 5.

10. *Catechism of the Catholic Church*, p. 547.

11. The Holy See, "Observations on the Universal Declaration on the Human Genome and Human Rights," November 11, 1997.

12. *Donum vitae. Instruction*, I, 2. Italics in original.

13. *Donum vitae. Instruction*, I, 2.

14. *Catechism of the Catholic Church*, p. 552.

15. *Catechism of the Catholic Church*, p. 549.

16. *Donum vitae. Instruction*, I, 4. Italics in original.

17. *Donum vitae. Instruction*, I, 4.

18. *Catechism of the Catholic Church*, p. 552.

19. *Catechism of the Catholic Church*, p. 552.

20. Pontifical Academy for Life, Plenary Assembly, "Concluding Document," Vatican, 1998.

21. John Paul II, "Address to the World Medical Association," Vatican, October 29, 1983.

22. Pontifical Academy for Life, "Concluding Communiqué on the 'Ethics of Biomedical Research. For a Christian Vision'," Vatican, February 24–26, 2003, p. 2. Italics in original.

23. "Prospects for Xenotransplantation," p. 10. Italics in original.

24. "Prospects for Xenotransplantation," p. 11. Italics in original.

25. "Prospects for Xenotransplantation," p. 11.

26. "Concluding Communiqué," p. 2. Italics in original.

27. *Donum vitae. Instruction*, I, 4. Italics in original.

28. *Catechism of the Catholic Church*, p. 549.

29. "Reflections on Cloning," p. 4.

30. *Catechism of the Catholic Church,* quoting *Donum Vitae. Instruction*, I, 6. Italics in original.

31. Pontifical Academy for Life, "Reflections on Cloning," Vatican, 1997, p. 6.

32. Cardinal William H. Keeler, chairman, Committee for Pro-Life Activities, "Letter to Congress," May 21, 2001.

33. The Holy See, "Intervention by the Holy See Delegation at the Special Committee of the 57th General Assembly of the United Nations on Human Embryonic Cloning, p. 2.

34. The Pontifical Academy for Life, "Declaration on the Production and the Scientific and Therapeutic Use of Human Embryonic Stem Cells," August 2000. para 3, p. 4.

35. "Declaration on the Production and the Scientific and Therapeutic Use of Human Embryonic Stem Cells," para, 5, p. 5.

36. Pontifical Academy for Life, Plenary Assembly, "Concluding Document," February 1997.

37. Joint Committee on Bioethical Issues of the Catholic Bishops [of Great Britain], "Genetic Intervention on Human Subjects," London, 1996, p. 33. Italics in original.

38. "Genetic Intervention on Human Subjects." Italics in original.

39. International Theological Commission [Catholic Church], "Communion and Stewardship: Human Persons Created in the Image of God," paragraph 90 (Vatican, 2002). Available at http://www.vatican.va/roman_curia/congregations/cfaith/cti_documents/rc_con_cfaith_doc_20040723_communion-stewardship_en.html (accessed May 7, 2007). Published with the permission of Joseph Cardinal Ratzinger, then the president of the Commission and now Pope Benedict XVI.

40. "Genetic Intervention on Human Subjects." Italics in original.

41. "Genetic Intervention on Human Subjects."

42. "Concluding Communiqué," p. 3.

43. John Paul II, Address to the Pontifical Academy of Sciences, L'Osservatore Romano (Eng. ed.), November 8, 1982. Cited in Richard Doerflinger, "Human Experimentation and the Sanctity of Life," USCCB Committee for Pro-Life Activities, June 2003, p. 6.

44. Thomas A. Shannon and Allan B. Wolter, OFM, "Reflections on the moral status of the pre-embryo," *Theological Studies* 52 (December 1990): 603–626.

45. For a further development of these ideas, see Thomas A. Shannon, "Cloning, uniqueness, and individuality," *Louvain Studies* (Winter 1994): 283–306.

46. *Donum vitae. Instruction*, I, 2.

# 4

# A Traditional Christian Reflection on Reengineering Human Nature

H. Tristram Engelhardt, Jr.

## Whose Christianity? Which Sense of the Traditional?

To draw coherent moral guidance from Christianity about applying genetic engineering to humans, one must first determine to which of the many Christianities one should turn and why. This determination involves taking a position as to what Christianity is. For example, one might conclude that Jehovah's Witnesses and the Christian Scientists do not hold representative Christian views regarding bioethical matters. Even an appeal to traditional Christianity to gain a more representative or at least more typical view[1] is beset by ambiguities. Such an appeal is surely contentious. Confronting the challenge of determining which Christianity should guide and why it should will commit one to taking a stance not only about the nature of Christianity, but about the character and significance of Christian tradition. Only in the light of what Christianity is and what Christianity knows can one determine what guidance Christianity can plausibly give regarding the use of our developing capacities in genetic engineering.

Taking a stand as to the nature of Christianity and as to what Christianity knows is provocative. Addressing the first issue will evoke considerable disputes among those who claim to be Christian, as well as with those who claim to know something about Christianity. On the one hand, one can draw on religious studies and the sociology of religion to determine which Christianity is typical in which respects by exploring the complex interplay of Christian institutions, practices, and cultural commitments. In this context, one would not need to presuppose the existence of natural kinds or ontologically secured boundaries separating the Christianities. Instead, one could attempt to frame different categories, classifications, and boundaries in the service of disclosing how different value commitments and ways of life are associated with different senses of Christianity and how these offer different insights into the moral costs and benefits of human genetic engineering. Classifications of Christianity and Christian tradition would in this context be instrumental. They would not be aimed at reflecting some deep reality. Instead, such classifications of different Christianities could be used to determine how these various Christianities in different fashions heuristically disclose various appreciations of the ethical, social, and public policy implications of genetic science and technologies. Claims of religious, metaphysical, realities would be eschewed.

On the other hand, one can approach the issue of deriving guidance from Christianity as a matter of seeking religious or metaphysical truth. That is, one can take seriously the claim that the church is the body of Christ in the world (Colossians 1:24). One would then understand the task of identifying that Christianity from which one should seek guidance as the task of identifying the true, mystical body of Christ. Unlike the first, more sociological-moral undertaking, which reduces the significance of Christianity to moral and cultural concerns,[2] this approach appreciates the plurality of Christianities as a sinful consequence of a departure from or failure to embrace truth. While the first approach is ecumenism friendly, the second approach is not. Any substantive position regarding the nature of Christianity will very likely collide with substantive ecumenical aspirations. That is, if it is understood that there is a truth of the matter as to what Christianity is and ought to be, then there is also a truth of the matter as to what Christianity is not and should not be, and therefore as to who should determine the use of genetic engineering.

Taking a stand as to which Christianity one should turn for guidance will evoke even more contention and debate because the implications will bear, not just against many Christians, but on religious claims generally. If some Christians know what others do not know in virtue of their privileged religious experience, it will follow that other religions do not know what they should know, or what could be known within a rightly ordered religious perspective. To say the least, such claims will have even more broadly nonecumenical consequences. Understood as a matter of cultural richness, claims in this regard concerning what religion has to offer may not be that provocative, even if they are unsettling for some (for example, consider the claim "Roman Catholicism has greater, historically grounded intellectual-cultural resources to speak to the question of the morality of human genetic engineering than do the Pentecostals"). However, those who take Christianity seriously as the body of Christ in the world will be asking a quite different question: "What does the church as the body of Christ in the world teach as the truth of the matter regarding the proper use of genetic engineering with humans, and which is that true church that teaches rightly?" To recognize the choice of a religion as disclosing matters of truth is currently countercultural. In particular, to take Christianity seriously as disclosing matters of truth regarding the deep nature of reality as well as the requirements for salvation will require facing the antiecumenical consequences of this recognition. At the very least, one will need to keep vividly in mind two important points: First, truth is cardinal to rightly ordered love; one must speak the truth with love about matters of ultimate meaning. Second, one's attitude toward those whom one is convinced know the truth only partially or not at all should be one of affirmative kindness, not of disrespect. One must approach all with love, but especially those mistaken about issues of ultimate importance.

The question then is what is added by the qualification "traditional" in speaking of Christianity. After all, if there is a church that is the body of Christ, why should one then characterize that church as traditional? Here the qualification traditional is in support of two goals. The first is to distinguish a Christianity that stands against posttraditional Christianity. We live in a period that experiences itself as liberating its institutions, including its religious institutions, from what many hold to be the misguiding and wrongly constraining commitments of the past. In this context, the characterization or traditional takes on a strongly negative valence in contrast to that which is considered progressive and liberating. One might think, for example, of those Christian sects that have claimed to ordain priestesses or bless homosexual unions. Traditional in the context of such Christian sects identifies a wrongly and oppressively constraining source for guidance embedded in a past that should be set aside. Of course, from the perspective of the traditional believer, post-traditional religious commitments involve a crucial rejection of truth and a distortion of reality. The traditional and the posttraditional stand to each other in robust opposition.

Since the goal of this chapter is to lay out the guidance Christianity can give regarding the proper use of human genetic engineering, it is important to choose the right Christianity. This in turn leads one to consider traditional versus posttraditional Christianity so as to identify a community with institutions and practices that are in continuity with an original Christianity whose patterns of thought, action, and belief are taken historically to define Christianity. A second and allied goal is to use a traditional view to identify not just a continuity of social patterns, moral attitudes, and ecclesiastical institutions sustained by customs handed down from the past, but to do so in the service of a frankly religious-metaphysical agenda of identifying that community which enjoys continuity in the truth through the Holy Spirit. This chapter thus begins first with the acceptance of Christ as the Son of the living God and the Messiah of Israel, as well as of his presence in the church as his body in the world. This reflects the crucial answer as to who Christ is, given by the apostle Peter as well as Martha, the sister of Lazarus whom Christ raised from the dead. To Christ's straightforward question, "Who do you say I am?" Peter answers, "You are the Christ, the Son of the living God" (Matthew 16:15). So, too, when Christ reminds Martha that he himself is "the resurrection and the life" (John 11:25), Martha responds, "You are the Christ, the Son of God" (John 11:27). Second, this essay takes seriously that the church is the body of Christ in the world (Colossians 1:24). It takes seriously the early Christian confessions of one, holy, catholic, and apostolic church as testimony that there is such a church, and that this church is Orthodox Christianity.

In the course of this essay, traditional Christianity will be understood as the Christianity that meets at least the following conditions, the first four of which are open to philosophical, historical, and sociological examination. These criteria will allow even non-Christians to appreciate the historical rootedness of what is advanced on behalf of traditional Christianity. Traditional Christianity

1. affirms without alteration or qualification the Nicean-Constantinopolitan Creed;
2. maintains an ecclesiology that is essentially the same as that which existed at the time of the council in Nicea (A.D. 325) and the First Council of Constantinople (A.D. 381);
3. exists as a community in historical continuity with the church that assembled at the Council of Nicea and the First Council of Constantinople (i.e., it takes seriously the ninth article of the creed: "I believe in one, holy, catholic, and apostolic church");
4. lives within an understanding of theology that is essentially the same as that existing at the time of the Council of Nicea and the First Council of Constantinople; and
5. lives in the Holy Spirit as the body of Christ in the world.

The church in the fourth century, the church as it emerged from persecution, is taken as the point of reference for identifying traditional Christianity because it is only in the fourth century after St. Constantine the Great (†337) when the threat of repression was largely gone (except for Julian the Apostate, who reigned A.D. 360–363) that the church could for the first time leave extensive records of its life. It is in this period that the church also enjoyed the presence and witness of St. Ambrose (A.D. 340–397), St. Anthony the Great, professor of the desert (A.D. 251–356), St. Athanasius (A.D. 295–373), St. Basil the Great (A.D. 329–379), St. Cyril of Jerusalem (A.D. 315–386), St. Ephraim the Syrian (A.D. 306–373), St. Gregory the Theologian (A.D. 329–390), St. John Chrysostom (A.D. 354–407), and St. Pachomius of Egypt (A.D. 286–346). This cloud of witnesses to the spirit, character, and life of the church supplies a robust picture and record of the Christian church and its judgments regarding a wide range of issues, including many bearing on medicine.

This observation does not collide with the circumstance that, in order for the church in the fourth century to be an exemplar of traditional Christianity, it must be in accord with the Christianity of Acts, the apostolic epistles, and the apostolic fathers. The difficulty is that these sources, with the exception of the letters of St. Ignatius of Antioch, provide only limited windows onto the ecclesial structure of the church. It is the church in the time of the Council of Nicea and surely the church by the time of the First Council of Constantinople of which we have a full picture. The age of the first two ecumenical councils also had the advantage of still being framed within the Semitic-Greco-Roman culture within which Christ preached and the apostles evangelized. It is a church that still easily thinks and lives within the commitments of a paradigm at one with the unbroken Semitic tradition that one finds in the writings of St. Isaac the Syrian of Nineveh (seventh century). It is an understanding of the church undistorted by the influences of the Frankish impact on the Roman papacy, the Scholastic intellectual synthesis of the western thirteenth century, the Renaissance, the Reformation, and the posttraditional consequences of the Second Vatican Council.[3]

Most important, this church appreciates theology, not as primarily an independent academic discipline, but instead as union with God achieved in prayer. "If you are a theologian, you will pray truly. And if you pray truly, you are a theologian."[4] In that experience of God, Christians can recognize that since "Jesus Christ is the same yesterday and today and forever" (Hebrews 13:8), so, too, they can recognize that the church is the same yesterday, today, and tomorrow. The church is nothing less than Christ's body "that is the Church" (Colossians 1:24). In this light, Tradition is encountered as the Holy Spirit sustaining Christianity's unity of right worship and right belief, which unity embraces only that which is "believed everywhere, always, and by all." Traditional Christianity thus affirms the criteria of universality, antiquity, and consent,[5] marking and maintaining the one, holy, catholic, and apostolic church in its unity of worship and belief.

The Christianity that lives in this experience of a unity and community over time and place of right worship and right belief, uniting itself to the age of the apostles and the fathers, is Orthodox Christianity. That is, traditional Christianity in this chapter is identified with Orthodox Christianity. Because it lives in the mind of the apostles and the fathers, its responses to questions regarding the proper use of human genetic engineering will be located in and drawn from resources already available in the first centuries of Christianity. It proceeds with a confidence that its truth unites past and future because the church is the body of Christ in the Holy Spirit and "Jesus Christ is the same yesterday, today, and tomorrow" (Hebrews 13:8). As a consequence, even apparently new challenges will be located in the moral framework and expectations of the experience of Christianity's first centuries. Here, the use of traditional indicates a robust unity of perspective and teaching over space and time.

## What Christianity Knows

United in true theology (i.e., in union with God), traditional Christianity has knowledge concerning the nature of the universe, the purpose of human life, and the content of morality, all of which is important for the proper use of human genetic engineering. Thus, what Christianity knows is unknown, partially known, or distortedly known by others. The knowledge that traditional Christianity possesses is essential for adequate orientation in the cosmos. In particular, traditional Christianity knows that

1. Jesus Christ is the long-awaited Messiah of Israel, who has been born, was crucified, and has risen from the dead and will come again to judge this world.
2. The history of the cosmos and of mankind stretches from the Creation through the Fall, incarnation, and redemption, and is on its way to restoration at the glorious Second Coming of Christ.
3. The good and the right can only be appreciated in terms of the holy, since the first and greatest of the commandments is that "you love the Lord your God with all your heart and with all your soul and with all your mind" (Matthew 22:37).
4. Natural law accounts as they have come to be framed fail to aim rightly at God in that they are inevitably distorted by the surrounding culture and the philosophical conceits of the times.[6] Such accounts attempt to find traces of God in nature, rather than to look through nature as through an icon so as to see God, while true moral theology is grounded in man's experience of the living and intervening God, who is in himself unknowable.[7]
5. God created humans in ontologically distinct sexes as male and female. "Haven't you read . . . that at the beginning the Creator 'made them male and female' (Genesis 1:27)" (Matthew 19:4).

6. The humanity taken on by Christ in the incarnation has been redeemed and, given the headship of both the first and second Adam, the daughters of Eve have likewise been redeemed. The dignity of humanity is only appreciated one-sidedly and incompletely apart from the Creation. The dignity of humanity is especially rightly appreciated through the incarnation, in that through the incarnation God became man "that we might be made God."[8]

Any account of the proper use of genetic engineering will need to be embedded in Christianity's rich knowledge of the meaning of the universe, human history, and the moral significance of human nature. Central to all reflections in this matter is the circumstance that God created human nature as good and appropriate, and that this very nature, albeit fallen, was taken on by Christ and redeemed.

It is important to note that the Christianity of the first millennium was unencumbered by the natural law theory that developed in the mid-thirteenth century. For the Christianity of the first millennium, natural law was not a normative structure understandable apart from rightly recognizing God.[9] Christian concerns with morality and metaphysics were located within a way of life directed toward the pursuit of holiness. The ecclesiology, the mysteries, the moral epistemology, and the metaphysics of this Christianity are integral to a religious way of life. They do not exist outside of that life with a critical power or authoritative status able to revise that which has been received. This is the case, even though traditional Christianity of the first millennium imported theological terminology from secular philosophies, and then employed it in discursive rational arguments in the service of apologetics and in disputes with those outside the faith. Traditional Christian theology and morality, which were framed within the embrace of the first millennium, were recognized as secured by an enduring experience of God grounded in grace (the uncreated energies of God). It is this experience of God (in western terms this would be characterized as mystical) that maintains the traditional Christian community of worship and belief over space and time.

As a result of this epistemological anchoring in a noetic experience of God, there is implicitly a distinction between theologians in the primary versus theologians in a secondary sense. Theologians in the primary sense are those who have noetic experience of God. They need not be, and are quite frequently not, academics.[10] Theologians in the secondary sense are those who are merely academic theologians who without experiencing God serve as translators of the experience of theologians in the first sense into the language of the general culture. Given the recognition of this distinction, theologians in the second sense (i.e., those who have merely discursive rational philosophical knowledge regarding God and morality) thus cannot bring into question what is given to the church by theologians in the first sense (i.e., those who have experience of God and his commandments). As a result of this circumstance, theology in the second sense cannot function as an independent moral-philosophical practice with critical reversionary authority, as occurs in many western Christianities, so as to bring theology in the first sense as well as traditional beliefs into question and to revise them. Given this understanding of theology, theological experts in the primary sense may not be found where western philosophy and theology would first think to look for them.

Those experts are true theologians who experience God's energies so that they do not simply know about God, they know God. This is the case because true theologians achieve their knowledge through a relationship with God achieved through rightly directed prayer. Traditional Christianity, Orthodox Christianity, continues to maintain that morality and theology are one with its experience of God. As the Ecumenical Patriarch Bartholomew I puts it,

> Therefore we do not engage in idle talk and discuss intellectual concepts which do not influence our lives. We discuss the essence of the Being Who truly is, to Whom we seek to become assimilated by the grace of God, and because of the inadequacy of human terms, we call this the image of the glory of the Lord. Based on this image, and in the likeness of this image, we become "partakers of the divine nature" [2 Peter 1:4]. We are truly changed, although "neither earth, nor voice, nor custom distinguish us from the rest of mankind." [To Diognetos 2, PG 2, 1173]
>
> This change, which is bestowed on us from the right hand of the Most High, remains hidden, secret and mystical to many. And thus, a life which is directed toward Him is called mystical. That which leads to divine grace are called mysteries. The entire change of both language and intellect is beyond comprehension and when directed by God leads to unspeakable mysteries.[11]

As a result, any analysis of the morality of human genetic engineering must be examined within the context of a religious life with God. In this context, there will not be independent moral rules or principles as external canons guiding the answers, but rather constraints and points of direction that are integral to religious life itself. What is offered in this chapter is at best theology in the second sense. Decisions regarding hard cases will in particular need to be referred to the true theologians, the holy fathers of the twentieth-first century.

## What the Christianity of the First Millennium Has to Say about the Appropriate Use of Human Genetic Engineering

Against this theological background, the question is how to engage the knowledge Christianity brings to the service of guiding societies, scientists, physicians, patients, and others in the development and proper use of human genetic engineering. One should note that such a search for answers and direction from religion already has a vague expression and resonance in secular society. Francis Fukuyama in an otherwise secular volume acknowledges that "religion provides the clearest grounds for objecting to the genetic engineering of human beings, so it is not surprising that much of the opposition to a variety of new reproductive technologies has come from people with religious convictions."[12] Fukuyama also remarks that parts of Asia, which are largely uninfluenced by Christian culture, have difficulty discerning what special moral issues could be at stake in the use of genetic engineering in humans.[13] Fukuyama wishes to use religion for the purposes of his public policy agendas, so as to limit the misuse of genetic technologies, all without recognizing religion as a cardinal source of moral and metaphysical knowledge and orientation.

A similar appreciation of the role that religion plays in highlighting issues to which a thoroughly secular culture is blind is made by Habermas, who nevertheless wishes to take the insights mined from religion and transmogrify them through a secular lens to make them generally available. "Those moral feelings which only religious language has as yet been able to give a sufficiently differentiated expression may find universal resonance once a salvaging formulation turns up for something almost forgotten, but implicitly missed. The mode for nondestructive secularization is translation."[14] In short, Habermas sees that there is a problem in providing substantive moral guidance in a high-technology culture, in that philosophy is insufficient to deliver the direction sought. To secular philosophy, it will appear as if there were no ultimate purpose to mankind or the universe. They will appear ultimately to come from nowhere, go to no place, and for no purpose, making it impossible to set ultimate directions and establish substantive constraints on the use of human genetic engineering. Like Fukuyama, Habermas recognizes that religion promises such direction. Yet, Habermas like Fukuyama then deconstructs the strength of religion, especially Christianity, which could have provided the needed guidance. He steps back from knowledge that can only be possessed by recognizing the relationship between everything in the cosmos and God.

If Christianity is what it claimed to be in the first millennium, namely, the very locus of God's revelation, then an answer to the question about the proper use of genetic engineering for humans requires entering into the paradigm or mind of traditional Christianity (i.e., the *phronema* of the fathers) so as to see how an answer can be found and understood. One needs at the very outset to know the significance of being human and what, if any, changes in human biological nature one should not make. Such guidance is what Christianity offers. Given traditional Christianity's appreciation of the presence of the Holy Spirit binding the church through its history, the church will turn to its past in order to address the problems of the present to avoid being misguided by the passions of the present. In order to give guidance regarding human genetic engineering, the office of a theologian in the second sense will here be to draw on guidance given by those who are theologians in the first sense, such as St. Basil the Great (A.D. 329–379).

One must begin with Christianity's traditional acceptance and affirmation of medicine and medical interventions. The church has taught that all things being equal, the use of such interventions is morally required. A synopsis of the traditional Christian view in these matters is provided by St. Basil the Great in question 55 of *The Long Rules*. There he responds to the question as to whether the use of medicine is morally acceptable. First, St. Basil affirms, "Each of the arts is God's gift to us, remedying the deficiencies of nature, as, for example, agriculture, since the produce which the earth bears of itself would not suffice to provide for our needs."[15] He then endorses medicine. "And, when we were commanded to return to the earth whence we had been taken and were united with the pain-ridden flesh doomed to destruction because of sin and, for the same reason, also subject to disease, the medical art was given to us to relieve the sick, in some degree at least."[16] Medicine, medical interventions, and surgical procedures are affirmed.

While affirming the propriety of using medicine and underscoring the obligation to employ it, St. Basil sets important limits. One must guard against turning the pursuit of health, longer life, or medical progress into what is tantamount to an idol. "Whatever requires an undue amount of thought or trouble or involves a large expenditure of effort and causes our whole life to revolve, as it were, around solicitude for the flesh must be avoided by Christians. Consequently, we must take great care to employ this medical art, if it should be necessary, not as making it wholly accountable for our state of health or illness, but as redounding to the glory of God and as a parallel to the care given the soul. In the event that medicine should fail to help, we should not place all hope for the relief of our distress in this art, but we should rest assured that He will not allow us to be tried above that which we are able to bear."[17] In short, on the one hand the use of medicine is endorsed, yet on the other hand it is placed within important constraints.

Drawing on these understandings, one can lay out a number of negative and positive conclusions with regard to the use of genetic engineering with humans. These conclusions must be understood as integral to a way of life aimed at holiness, where the good is not reduced to the holy, and the good apart from the holy is always perverse. Their sense and force are embedded in a wholehearted pursuit of union with God. First, five negative, constraining conclusions can be articulated.

1. The use of genetic engineering, whether somatic or germline, should not become an all-consuming cultural, societal, communal, or individual undertaking. Any endeavor is forbidden that places the solicitude of the flesh first and foremost. Turning this life into an idol, suggesting that one can forget that the prize is not health or indefinite extension of this worldly existence, is forbidden. Thus, the pursuit and employment of genetic engineering may not be used to distract from the primary goal of humans: union with God.

2. As an extension of the first negative constraint, children must always be ascetically accepted as gifts of God, so that any use of human genetic engineering or any other medical intervention will not be regarded as the creation of an object or the design of a child, but as an engagement with God's help through technology to act to benefit one's children.[18]

3. One may not alter the character of humans as male and female. The ontological expression of humans in two sexes is established in paradise and affirmed in the New Testament (Genesis 1:27; Matthew 19:4). It is integral to the struggle for salvation.

4. One may not alter the general character of human biological nature and the human body so that the body of humans becomes different from the body assumed by Christ, who in the incarnation took on our form. The general human form and character are sanctified both by creation and by the incarnation.

5. No destruction of an embryo should occur in the process of human genetic engineering. The church from its beginning understood that the destruction of early human life is wrong, whether or not ensoulment has taken place (St. Basil the Great, Letter 188).

These five constraining conclusions are expressions of a general concern about how one should live so as to come into union with God. This approach to the question of the proper goals for human genetic engineering radically relativizes the project of human genetic engineering. What medicine can promise for earthly health and what human genetic engineering can offer in the way of improving the human condition pale in comparison with what Christianity offers: immortality and union with God. In this light, the undertaking of the transhumanists underestimates the radical future open to humans. They fail to recognize that the truly transhuman project is that achievable through Christ's incarnation, namely, theosis.[19]

Three positive conclusions can now be articulated. They acknowledge that the application of genetic engineering to humans can be approved where there is a legitimate therapeutic goal or even the prospect of enhancing, that is, restoring, human biology broken after the Fall. Of course, one is still required to act within the five foregoing constraints, as well as other constraints of the Christian life (e.g., one should not steal the resources needed to pay for treatment). In addition, there must be consent of the participants and good grounds to hold that there will be more benefit than harm in that all treatment carries risk.

1. Curative medical interventions are not only permissible but obligatory as long as they do not violate the foregoing five constraints or some other general moral prohibition (e.g., killing another to acquire his organs). As a consequence, the technological interventions of somatic and germline genetic engineering aimed at curing human disease would not be categorically forbidden, but indeed in some cases would be recognized as obligatory.
2. Forms of genetic enhancement that increase resistance to disease, disability, and early death within the aforementioned constraints are acceptable insofar as they address the harm done to human biology that is due to the Fall, as long as they do not attempt to alter the biological character of humans. According to Scripture (Genesis 5:25), the life of Methuselah spanned hundreds of years.[20]
3. Within the negative constraints articulated here, both somatic and germline genetic engineering can be used to alleviate human suffering and to increase human resistance to disease, disability, and early death. However, it is important that these interventions never produce a so-called posthuman nature, but rather support the humanity created and blessed by God. One must recall that humans already rather routinely receive not only transplants, but artificial implants (e.g., heart valves and plastic lenses after cataract surgery), all of which are ensouled by the person receiving them. Humans must recognizably continue to possess the biological humanity taken on by Christ.

These three positive conclusions offer a generally friendly response to the core aspirations of human genetic engineering: genetic engineering is *ceteris paribus* to be approved in the pursuit of ameliorating disease and improving health.

In summary, although traditional Christianity has concerns that limit and direct human genetic engineering, concerns that it does not share with the secular culture, these do not create a categorical prohibition in principle against such technology. *Pace* many secular moralists, genetic engineering of the human germline can be endorsed within constraints.

Within these constraints, it is not just that there would be no grounds to prohibit the use of such technologies, but indeed there would be strong grounds in favor of using genetic engineering to remove dangerous mutations that are the basis of disease and disability, not only in the persons afflicted, but from the germline itself. For example, one should *ceteris paribus* use human genetic engineering to cure type I diabetes, not only in those who are ill, but in their descendants. Seen in this light, human genetic engineering, if used within the constraints outlined, can be thought of as a special form of microsurgery.

## A Concluding Puzzle and Postscript: Why Some Secular Moralists Have Objections in Principle Against All Germline Genetic Engineering

Against the backdrop of the traditional Christian position regarding the use of genetic engineering in humans, how can one explain the often categorical secular prohibition of human germline genetic engineering? Of course, there are legitimate grounds for worrying about misuse as well as untoward social consequences. There are "slippery slope" hesitations of the form that engaging in even legitimate human genetic engineering may make illegitimate uses seem more acceptable and therefore make abuses more likely. There are also legitimate reasons to balance benefits and harms properly. However, none of these considerations will produce a categorical prohibition, a prohibition in principle. Consider, for example, the Convention for the Protection of Human Rights and Dignity of the Human Being with regard to the Application of Biology and Medicine: Convention on Human Rights and Biomedicine (Oviedo, April 4, 1997). "Article 13—Interventions on the human genome. An intervention seeking to modify the human genome may only be undertaken for preventive, diagnostic or therapeutic purposes and only if its aim is not to introduce any modification in the genome of any descendants."[21] This article absolutely and categorically forbids human germline genetic engineering. Yet there do not appear to be general secular grounds for such a prohibition.[22]

An explanation for this phenomenon lies in the displacement of transcendent concerns.[23] In a culture once Christian and now secular, moral sentiments remain from that past, albeit disconnected from their previously supporting moral and metaphysical framework. They persist as free-standing intuitions, hesitations, and taboos. As a result, there is often a lingering sense that there is something wrong about a particular undertaking, a feeling that one ought to have a moral concern, but an inability to recognize rightly why and how to focus that concern. This phenomenon has been recognized at least in part by Alasdair MacIntyre.[24] However, it is not just that one is left with intuitions without sufficient metaphysical scaffolding, which become mere taboos. In addition, these moral intuitions and sentiments often attach themselves to actions in a distorted fashion. Thus there is a sense that one should approach human germline genetic engineering with moral concern, but the concern is often difficult to focus.

Given the displacement of transcendent concerns, one often wants to say what can no longer be said. This is illustrated by a discourse based on human dignity when it is used to establish a special constraint on human genetic engineering. The cardinal difficulty is that there is no secular basis for venerating humanity as we find it. Human nature, regarded merely as a contingent, secular given, that is, understood in purely secular terms, is a surd product of spontaneous mutation, selective pressures, genetic drift, and various random catastrophes. It is a nature that could have been otherwise and whose particularity in itself has no normative standing. It is Christianity that discloses the significance of being human through the truth of the Creation and the incarnation, thus providing orientation in the cosmos and a cardinal ground for appreciating the radical importance of humanity. It is Christianity that discloses the true dignity of humanity in the Creation, but much more importantly through the incarnation. However, only when one is rightly oriented to the Creator will one know rightly what this dignity involves (Romans 1:22–32). Wrong worship and wrong belief will deform one's moral vision and sensibility.

## Notes

1. "Typical" is used in the sense of an exemplar: laying out the characteristic features of Christianity, not necessarily those features most common among the various Christian sects.

2. Immanuel Kant provides a classical presentation of the Enlightenment's attempt to reduce religion to its moral significance or influence. This project would (1) reduce the holy to the good and (2) establish one universal and rational religion. As Kant puts it, "[T]he sacred narrative, which is employed solely on behalf of ecclesiastical faith, can have and, taken by itself, ought to have absolutely no influence upon the adoption of moral maxims, and since it is given to ecclesiastical faith only for the vivid presentation of its true object (virtue striving toward holiness), it follows that this narrative must at all times be taught and expounded in the interests of morality; and yet (because the common man especially has an enduring propensity within him to sink into passive belief) it must be inculcated painstakingly and repeatedly that true religion is to consist not in the knowing or considering of what God does or has done for our salvation but in what we must do to become worthy of it." [Immanuel Kant, *Religion Within the Limits of Reason Alone*, trans. T. M. Greene and H. H. Hudson (New York: Harper, 1960), p. 123, AK VI, 132f.]

3. The Roman Catholic Second Vatican Council (1962–1965) produced a radical and thoroughgoing recasting of the life of its faithful by (1) altering the character of worship, (2) removing nearly completely any vestiges of traditional Christian ascetic commitments, and (3) creating a sense that the doctrines and life of the church should be accommodated to the concerns of the secular culture. See, for example, Michael Davies, *Pope Paul's New Mass* (Dickinson, TX: Angelus Press, 1980); *Pope John's Council* (Kansas City, MO: Angelus Press, 1977); and James F. Wathen, *The Great Sacrilege* (Rockford, IL: Tan Books, 1971). See also H. T. Engelhardt, Jr., *The Foundations of Christian Bioethics* (Lisse, Netherlands: Swets & Zeitlinger, 2000), pp. 53–55. These changes were associated with a rapid exit of priests and nuns, as well as a dramatic drop in the number of vocations. See, for example, Kenneth C. Jones, *Index of Leading Catholic Indicators* (Fort Collins, CO: Roman Catholic Books, 2003).

4. Evagrios the Solitary (A.D. 345–399), "On Prayer," in Sts. Nikodimos and Makarios, *The Philokalia*, trans. and eds. G. E. H. Palmer, Philip Sherrard, and Kallistos Ware (Boston: Faber and Faber, 1988), vol. 1, p. 62. At stake is a crucial distinction between the original theology of Christianity and that which emerged in the West. The former understood theology, as did St. Symeon the New Theologian (A.D. 949–1022) and St. Gregory Palamas (A.D. 1296–1359). They emphasize a noetic experience of God rather than a discursive, philosophical undertaking. See, for example, Hierotheos Vlachos, *The Mind of the Orthodox Church*, trans. Esther Williams (Levadia, Greece: Birth of the Theotokos Monastery, 1998). In the West, theology was no longer understood as a successful relationship to God, but it became a third thing, an academic practice mediating the relationship of God and man.

5. St. Vincent of Lerins, *A Commonitory* II,4,6, in *Nicene and Post-Nicene Fathers*, second series, Philip Schaff and Henry Wace, eds. (Peabody, MA: Hendrickson Publishers, 1994), vol. 11, p. 132. The Orthodox Church, for example, recognizes the church's presence in an unchanging understanding of

right worship and right belief grounded in and secured by the presence of the Holy Spirit, whose presence is sacred tradition. As St. Silouan the Athonite (A.D. 1866–1938) puts it, "Sacred Tradition, as the eternal and immutable dwelling of the Holy Spirit in the Church, lies at the very root of her being, and so encompasses her life that even the Scriptures themselves come to be but one of its forms. Thus, were the Church to be deprived of Tradition she would cease to be what she is, for the ministry of the New Testament is the ministry of the Spirit 'written not with ink, but with the Spirit of the living God; not in tables of stone, but in fleshy tables of the heart.' Suppose that for some reason the Church were to be bereft of all her liturgical books, of the Old and New Testaments, the works of the holy Fathers—what would happen? Sacred Tradition would restore the Scriptures, not word for word, perhaps–the verbal form must be different—but in essence the new Scriptures would be the expression of that same 'faith which was once delivered unto the saints'. They would be the expression of the one and only Holy Spirit continuously active in the Church, her foundation, and her very substance." [Archimandrite Sophrony, *The Monk of Mount Athos: Staretz Silouan 1866–1938*, trans. Rosemary Edmonds (Crestwood, NY: St. Vladimir's Press, 1975), pp. 54–55.]

6. As Hegel appreciates, the dominant culture is framed by the categories it embraces so that as those categories change the culture itself changes. "All cultural change reduces itself to a difference of categories. All revolutions, whether in the sciences or world history, occur merely because spirit has changed its categories in order to understand and examine what belongs to it, in order to possess and grasp itself in a truer, deeper, more intimate and unified manner." [G. W. F. Hegel, *Hegel's Philosophy of Nature*, ed. and trans. M. J. Petry (London: George Allen and Unwin, 1970), § 246 Zusatz, vol. 1, p. 202.] Since there is nothing outside of the realm of being for thought and thought for being, being changes as the categories change.

7. The traditional Judeo-Christian encounter is with the God who commands, who must be obeyed, and whose requirements cannot be reduced to human natural law and/or philosophical moral requirements (i.e., the thoughts and ways of God are not the thoughts and ways of man—Isaiah 55:8). One remarkable illustration is the contrast between what the Torah in its 613 laws requires of Jews and what the covenant with Noah and his sons requires. "Seven precepts were the sons of Noah commanded: social laws; to refrain from blasphemy; idolatry; adultery; bloodshed; robbery; and eating flesh cut from a living animal. R. Hanania b. Gamaliel said: Also not to partake of the blood drawn from a living animal. R. Hidka added emasculation. R. Simeon added sorcery. . . . R. Eleazar added the forbidden mixture [in plants and animals]" (*Sanhedrin* 56a–b).

8. St. Athanasius, "De incarnatione verbi dei" §54.3, in *Nicene and Post-Nicene Fathers*, Schaff and Wace, eds., vol. 4, p. 65.

9. Christianity has traditionally recognized an integral connection of morality and religious life, especially between morality and right worship. This connection is emphasized by St. Paul in Romans 1:18–32. It should be underscored that St.

John Chrysostom (A.D. 354–407) in his commentary on Romans 2:10–16 considers that only righteous Gentiles such as Melchizedek were able correctly to understand moral obligations. "But by Greeks he [St. Paul] here means not them that worshipped idols, but them that adored God, that obeyed the law of nature, that strictly kept all things, save the Jewish observances, which contribute to piety, such as were Melchizedek and his, such as was Job, such as were the Ninevites, such as was Cornelius" "Homily V on Romans I.28, V.10," in Philip Schaff, ed., *Nicene and Post-Nicene Fathers*, first series (Peabody, MA: Hendrickson Publishers, 1994), vol. 11, p. 363.

10. For examples of twentieth-century theologians in this primary sense, see St. John of San Francisco (A.D. 1894–1966), Elder Joseph the Hesychast, the Cave-Dweller of the Holy Mountain (A.D. 1895–1959), Elder Paisios of Romania (†1993), Elder Paisios of Mt. Athos (A.D. 1924–1994), Elder Porphyrios (A.D. 1906–1991), St. Silouan the Athonite (A.D. 1866–1938), and Archimandrite Sophrony (A.D. 1896–1993).

11. Patriarch Bartholomew, "Joyful Light," speech delivered on October 21, 1997, at Georgetown University, Washington, DC, p. 3.

12. Francis Fukuyama, *Our Posthuman Future* (New York: Farrar, Straus and Giroux, 2002), p. 88.

13. Fukuyama, *Posthuman Future*, p. 192.

14. Jürgen Habermas, *The Future of Human Nature* (Cambridge, UK; Polity Press, 2003), p. 114.

15. St. Basil the Great, *The Long Rules*, in *Ascetical Works*, trans. Sister Monica Wagner (Washington, DC: Catholic University of America Press, 1962), p. 330.

16. *The Long Rules*, p. 331.

17. *The Long Rules,* pp. 331–332.

18. Would-be parents appropriately live healthy lives, take vitamins, etc., without thereby rendering the procreation of their children into an illicit designing of children. The difference will crucially be determined by the life within which procreation is nested.

19. For an account of theosis, which is nothing more than extended glosses on St. Athanasius the Great's (A.D. 295–373) remark that God became man so that men could become gods by grace, see Georgios I. Mantzaridis, *The Deification of Man* (Crestwood, NY: St. Vladimir's Seminary Press, 1984), and Panayiotis Nellas, *Deification in Christ* (Crestwood, NY: St. Vladimir's Seminary Press, 1987).

20. The distinction between genetic engineering aimed at treatment and that done for enhancement has proved notoriously difficult to express. For an overview of some of the challenges, see Emmanuel Agius and Salvino Busuttil eds., *Germ-Line Intervention and our Responsibilities to Future Generations* (Dordrecht: Kluwer Academic Publishers, 1998).

21. Council of Europe, Convention for the Protection of Human Rights and Dignity of the Human Being with regard to the Application of Biology and Medicine: Convention on Human Rights and Biomedicine (Strasbourg, France: April 4, 1997), chapter iv, article 13.

22. For an exploration of the inability of secular moral reflection to set limits to human genetic germline engineering, see H. T. Engelhardt, Jr., *The Foundations of Bioethics*, 2nd ed. (New York: Oxford University Press, 1996), especially pp. 411–418.

23. For a more detailed examination of the phenomenon of the displacement of Christian moral insight from a Christian framework to a secular context where Christian insights survive as "moral intuitions" and taboos, see H. T. Engelhardt, Jr., "Die menschliche Natur—Kann sie Leitfaden menschlichen Handelns sein? Reflexionen über die gentechnische Veränderung des Menschen," in *Die menschliche Natur*, Kurt Bayertz, ed. (Padernborn, Germany: Mentis Verlag, 2005), pp. 32–51.

24. Alasdair C. MacIntyre, *Whose Justice, Which Rationality* (South Bend, IN: University of Notre Dame Press, 1989).

# 5

# Germline Gene Modification and the Human Condition before God

Nigel M. de S. Cameron and Amy Michelle DeBaets

Genetic intervention in human beings, heralded a generation ago as the source of untold medical benefits and much moral consternation, has so far proved a disappointment. Despite the vast investment of the Human Genome Project and the monitoring apparatus of the Recombinant DNA Advisory Committee, therapeutic genetic interventions have yet to attain other than experimental status in clinical trials and even there have encountered serious setbacks.[1] The human genome, possessed of many fewer genes than had been anticipated, has proved a more intractable subject than early advocates of gene therapy had expected. Meanwhile, our new genetic knowledge has granted us far more diagnostic information than we can use. Newborns, fetuses, and in vitro embryos are subjected to batteries of tests that can identify hundreds of genetically based diseases.[2] Moreover, genetic traits are more complex than we had imagined, so any interventions to combat them will need to be proportionately sophisticated. The genetic reductionism that has inevitably accompanied our new knowledge, and its public dissemination, has been qualified by a growing awareness of the complexity of the relationship between our genes and ourselves.

One by-product of this slow progress in developing genetic interventions has been to postpone the question that lies behind this chapter. Had there been rapid development of genetic therapies, the question of inheritable interventions would soon have loomed large. Since, in essence, therapeutic genetic interventions have stalled, the discussion of their application beyond the narrowly medical model of therapy to the individual has hardly begun to find traction, either in the public mind or in the mind of the church. Yet the questions are fundamental. Somatic cell interventions may be therapeutic or may lead to enhancements in the individual; germline interventions may lead to either enhancements or therapy in future generations. They go to the core of our vision of what it means to be human, and raise questions regarding the proper role of technology in our exercise of stewardship of ourselves and, in the case of germline interventions, our children and our children's children.

What is more, we need to set potential developments in human genetics within the broader context of the full range of the new technologies and their impact on human nature. The National Science Foundation's (NSF) National Nanotechnology Initiative has done just that. In a series of conferences the NSF has developed the idea of converging technologies, in which biotechnology is brought together with nanotechnology, information technology, and cognitive science (nano-bio-info-cogno, or NBIC), for the putative purpose of "improving human performance."[3] While that phrase is in itself capable of various interpretations, the welcome accorded the process by transhumanists, committed to the reengineering of human nature into, ultimately, something else, has drawn attention to the fact that the prospect for improving humans is no longer wedded to genetics. Indeed, it has been suggested that nanotechnology itself will be the driver of such an enterprise. Meanwhile, the discussion of the ethics, theology, and public policy of genetic interventions in the germline has been conducted almost entirely in isolation from these wider developments in technology.[4]

So the technological context of this discussion is multifaceted. While the clinical applications of genetics have lagged, other technologies with relate potential are rapidly under development and beginning to serve as a sharper focus for the question of enhancement.

In tandem with advancements in these technologies, the public policy framework within which they may proceed has begun to develop. Both the UNESCO Declaration on the Human Genome and the European Convention on Human Rights and Biomedicine specifically address the question of germline interventions. The UNESCO Declaration states that germline interventions may be contrary to human dignity. The European Convention goes further and prohibits germline interventions. Both of these documents were completed in 1997. In 2005, the United Nations General Assembly completed its Declaration on Human Cloning that in addition to seeking to prohibit human cloning for any purpose, refers in general terms to genetic interventions that are contrary to human dignity. In the fall of 2005 UNESCO completed its Universal Declaration on Bioethics and Human Rights. Each of these documents will have the effect of setting the pace for domestic legislation in jurisdictions around the world.

## The Options for Human Genetics

Two sets of distinctions are commonly made in setting out the options for human genetic interventions. The first distinction refers to the cells to be affected and their relation either to the individual or his or her reproductive line. Genetic interventions may therefore be somatic, in that the cells to be affected are body cells rather than germline (reproductive) cells. This is the focus of most current research in human genetics. In somatic interventions, the genes of an existing individual are manipulated so that change is effected in that individual, but any such changes are not passed down to subsequent generations because they do not affect the reproductive cells. By contrast, germline interventions are performed on gametes (egg or sperm) or early embryos so that the individual's total genetic structure is altered and the genetic modifications are passed down to subsequent generations.

The second logical distinction refers to the purpose for which the intervention is made—whether that of therapy or enhancement. In this model, therapy is understood as an intervention with the goal of treating a disease or other genetic malfunction, while enhancement has the goal of adding capabilities or in other ways improving the condition of a person in whom the trait(s) being modified are already within the normal range. This distinction, while important in principle, in practice is not as useful as it might seem. In particular, it is hard to arrive at a definition of normality. Interventions that are plainly therapeutic for A (say, making a boy who is 4 feet 6 inches through some genetic defect or hormonal imbalance a foot taller so he is not embarrassed) become enhancements for B (making a boy who just happens to be 5 feet 6 inches into one who is 6 feet 6 inches, so he can play basketball). What constitutes a therapeutic intervention in one person would constitute enhancement in another. The difficulty is lessened if traits such as height, baldness, and obesity are excluded, but it does not disappear.

While it may be some time before these genetic options become possible, we experience them already in the use of human growth hormone. Indeed, one of the keys to an informed discussion of the potential of human genetics lies in the illustrations offered by nongenetic interventions, in which enhancements (especially in plastic surgery) are already common. Granted that we already make hormonal and surgical corrections to physical appearance, what notions of normal are we using?

The same problem in setting objective criteria in distinguishing normal height from deficient height arises with a wide variety of other traits, most of which are not strictly or simply genetically determined (such as intelligence, capability in music or sports, or moral sensibility). Since the question of enhancement is central to the germline debate, these examples from outside genetics are helpful in two crucial respects: They show the limitations of the therapy–enhancement distinction and they also show its importance.

## Toward a Theology of Medicine[5]

The debate about the proper uses of human genetics is in fact a debate not about science, or technology, or even medicine, but about anthropology—the nature of human being. What is human nature? Is it in our hands to do with as we choose, or is it given to us? For Christians, the anthropological question is not difficult to resolve. The Judeo-Christian view, which has so deeply influenced the western tradition, and the lineaments of which, through their secularization in the Enlightenment of the eighteenth century, have set the assumptions of the modern world, has been unambiguous that our idea of human nature is grounded in our being made in the image of God. This is then interpreted in the context of the incarnation of Jesus Christ, as God took on our humanity fully and perfectly, giving us a model of our nature.

With the entrance of sin into human life came all disease, suffering, and death. They may have no direct connection with the sins of an individual, but because of sin in the world, the natural state of human life is an unnatural one of disease and death. If disease and death have a fundamentally moral cause in sin, they have also a moral cure in redemption.

So just as all disease stems from sin, all healing stems from the work of Christ. His healing work was first seen in the context of his earthly ministry, in which he restored to health and wholeness people who were caught in a variety of conditions of disease and death. This ministry of healing was a foreshadowing of the work of redemption and healing accomplished for all on the cross. Just as the earthly ministry of healing looked forward to the cross, so also Christ's resurrection and overcoming of sin, disease, and death looks forward to the eschatological resurrection in which disease and death will be finally and forever overcome for all humanity. The Christian task of healing then, seeks to follow Christ's ministry of healing in response to the conditions of sin and death. We must understand and acknowledge that this ministry is both necessary and one that cannot be fully accomplished in this life. Rather, it is a witness to the eschatological hope we have in Jesus Christ and the coming of the new heaven and the new earth, in which sin and death will be no more. We recognize our limitations and rely on the providence of God, knowing that all of our efforts will be but a pale foreshadowing of what is to come. We may then also set medical healing and the healing that may come from prayer together as two sides of the one coin as in this life we anticipate in small ways the overcoming of death itself in the life to come.

Just as we cannot assume a causal relationship between sin in one's life and disease, so also we deny that virtue in one's life will correlate with health or prosperity. What health we have is a grace from God to be used to further the health of others, an inbreaking of the reign of God into the world to point us to Christ in eschatological anticipation. Likewise our task of healing, whether as physicians, clergy, or others in the caring and healing arts, is a calling to a ministry of grace for the whole person, a working toward restoration of health and wholeness that looks to the culmination and completion of health and wholeness, the overcoming of all suffering and death by Jesus Christ.

The effects of the cross are as cosmic as those of the Fall. The "whole creation" is still "groaning and travailing" as it waits with longing for the final purpose of redemption: the "adoption as sons, the redemption of our bodies" (Romans 8:23). The proper context of healing lies in the entire undoing of the ill effects of the Fall, as our bodies are redeemed at the resurrection of the dead.

## Creation in the *Imago Dei*

As we understand from Genesis, human beings were created in the image of God (1:26ff). The bearing of this image is inclusive of all people, regardless of gender, religion, age, or any other status. The image is species-specific, creating men and women in God's own likeness. As God's special creatures in his image, we have a special relationship of stewardship over the rest of creation. Genesis 1 declares that humankind was given dominion over the other creatures made by God—the beasts, birds, and fish—but not over one another. This creation of human beings in the image of God and our relationship of stewardship within creation are key as we explicate the implications of this doctrine in light of the incarnation of Christ.

As people created in the image of God, God has made us, and we have not made ourselves. God called the creation of humanity "good," including all people, regardless of their genetic "superiority" or lack thereof. Some theologians look at the dominion given to humanity in Genesis 1 and find in it a vocational calling that human beings function as "created co-creators."[6] A report of the Panel on Bioethical Concerns of the National Council of Churches of Christ, USA stated, "Dominion carries with it a concept of custody, of stewardship, of being responsible for, of caring for all creation. Therefore, we are called to live in harmony with all creation, including humankind, and to participate with the Creator in the fulfillment of creation."[7] It may easily and rightly be argued that this status of stewardship carries with it certain ongoing creative responsibilities as a significant part of our relationship as humans to the rest of creation, but dominion is explicitly not given to human beings over one another. Dominion over humanity still properly belongs to God alone, and so the argument that our responsibilities of dominion include the genetic specification and determination of one another falls apart.

The implications of the image of God in humanity in Genesis 1 have been explored throughout Jewish and Christian history. What can be known with little debate, though, is that this identification of the image of God with humanity is coextensive with human life itself, that the image is known fully and perfectly in Christ, and that the image of God in humanity gives to humanity a special relational and moral status in which human beings are to be specially protected as we live out our lives in the loving covenant of God's grace. Human beings were called from among the creatures of God to lives of unique responsibility, of moral awareness, of rational choices. Even as we are called to lovingly care for and protect all of creation, so much more are we to care for and protect those whom God has chosen to live in his image, the people among whom he became incarnate in order to save.

Even as the good creation of God, we are also fallen and sinful people, and we have no way to become perfect on our own. Our awareness of the moral cause of our disease and, finally, death will always temper our confidence in the technological interventions of medicine. Because technology cannot cure our sinfulness so no genetic manipulation will grant us moral perfection. It is only in Christ that we may have the hope of restoration to who we were created to be, and that will not in this life be perfected. We must care for one another, providing healing and restoration in faithful following of Christ's ministry of healing as we await the day when all healing will be accomplished and sin and death overcome. It is this eschatological perspective that drives our ministry and our hope. We are given by God a unique status as guardians of the creation, but this allows us power within limits, and does not give us dominion over one another.

## The Lewis Paradox

That takes us to a key concept that sheds unique light on the significance for the human condition of every technological intervention that is reflexive and not merely therapeutic, that is, that turns technology on our own selves and seeks to determine who we are. It is precisely at this point that the fact that dominion is given to us over the rest of creation, but not ourselves, becomes of unique relevance. It is also here that the co-creator idea, which in effect seeks to subvert that distinction and give us a share in godlike dominion over even our own selves, becomes so problematic.

In his famous essay, “The Abolition of Man,” an early version of which was first published in 1943, C. S. Lewis, the English literary scholar and lay theologian addressed from afar the coming challenges of human genetics and related technologies. His essay opens by noting the prima facie appeal of these new technologies with a poignant quotation from John Bunyan’s *Pilgrim’s Progress*: “It came burning hot into my mind, whatever he said and however he flattered, when he got me home to his house, he would sell me for a slave.”[8] That, in embryo, is Lewis’s percipient response to the prospect of the reflexive technologies and what lies beyond.

His argument opens with a consideration of the fact that all technology, which is said to extend the power of the human race, is in fact a means of extending the power of “some men over other men.” He instances the radio and the airplane as typical products of technology, which like all other consumer items, can be bought by some, not afforded by others, and could be withheld by some from others who have the resources to buy. In light of the pervasive influence of eugenic thinking and practice in the United States and the United Kingdom as well as Germany, in which enforced sterilization was widely employed for selective breeding purposes, Lewis builds his argument on the contraceptive and sterilization technologies of the early twentieth century even as he anticipates those of the twenty-first. As a result, he continues, “From this point of view, what we call Man’s power over Nature turns out to be a power exercised by some men over other men with Nature as its instrument.” He hastens to add that while it can be easily said that “men have hitherto used badly, and against their fellows, the powers that science has given them,” that is not his point. He is not addressing “particular corruptions and abuses which an increase of moral virtue would cure,” but rather “what the thing called ‘Man’s power over Nature’ must always and essentially be,” for “All long-term exercises of power, especially in breeding, must mean the power of earlier generations over later ones.”[9]

What Lewis is drawing attention to here is, as it were, the biological equivalent of what in another field is termed intergenerational economics. In the nature of the case, the genetic accounting is of a yet higher level of significance than economic relationships that run through time, although the principle is the same: the impact of one generation's decisions on subsequent generations. So Lewis states: "We must picture the race extended through time from the date of its emergence to that of its extinction. Each generation exercises power over its successors: and each, in so far as it modifies the environment bequeathed to it and rebels against tradition, resists and limits the power of its predecessors." There can be no increase in power on Man's side. "Each new power won *by* man is a power *over* man as well. Each advance leaves him weaker as well as stronger. In every victory, besides the general who triumphs, he is a prisoner who follows the triumphal car. . . . *Human* nature will be the last part of Nature to surrender to Man. The battle will then be won. We shall have 'taken the thread of life out of the hand of Clotho' and be henceforth free to make our species whatever we wish it to be. The battle will indeed be won. But who, precisely, will have won it?" Because "the power of Man to make himself what he pleases means, as we have seen, the power of some men to make other men what *they* please. . . . Man's final conquest has proved to be the abolition of Man."[10]

Lewis's analysis is directed at the possibility of germline (inheritable) genetic interventions, yet his twofold thesis is also of wider application. First he sets out the fundamental problem of biotechnology and its affiliates as a vast challenge that must be addressed and second he frames its significance precisely in the context of anthropology. While his argument uses public language, his starting point is the Christian understanding of what it means to be human, an understanding built deep into the western cultural tradition.

## The Case for Germline Interventions

The general debate over germline interventions is discussed most fully by LeRoy Walters and Julie Gage Palmer in their book *The Ethics of Human Gene Therapy*. They detail both the technical and ethical issues involved in germline engineering. In examining the technical issues, Walters and Palmer admit that significant technical barriers currently exist that prevent the possibility of safe and successful germline interventions. They then consider arguments in favor of and opposing the use of germline interventions in human beings; both believe that, in theory, germline interventions could be performed ethically. They admit, however, that this would require a "perfect world" scenario, in which all technical limitations and risks have been overcome and the interventions could be performed safely, effectively, and relatively inexpensively. This scenario clearly does not exist at present, so the arguments given could at best be persuasive in theory and dependent upon the resolution of ethical issues with the processes involved in overcoming the technical difficulties themselves.

Walters and Palmer set out key reasons typically given in favor of germline interventions in human beings. They argue that germline interventions "may be the only way to prevent damage to particular biological individuals when that damage is caused by certain kinds of genetic defects."[11] They claim that parents might wish to argue that they would like to spare their children from having to undergo somatic genetic interventions or other standard medical treatments or to avoid having those children face difficult decisions regarding passing on their own genes. They argue that in the long term, germline interventions would be less costly than somatic interventions and that researchers ought to have the freedom to explore new modes of treating and preventing disease. Finally, they argue that germline engineering could protect the lives of individuals with disabilities from the alternative choice of selective abortion.

These arguments have been answered in several ways. For one thing, the safety issues are highly significant and have been highlighted by problems already faced in more modest, somatic gene therapy clinical trials, symbolized in the tragic death of Jesse Gelsinger, who died in 1999 during one such trial. In the case of germline interventions, unanticipated negative effects would affect not only one person but all of that person's progeny. Moreover, such interventions, whether for treatment or enhancement, will likely always be relatively expensive, so that only the wealthy will be able to afford them; so genetic disease would not be eradicated on a public health level, but for privileged families. These interventions will only be developed at the cost of programs of embryo and fetal experimentation. Indeed, the perfection of the techniques of germline intervention will require experimentation on fetuses and children who would be used as human guinea pigs for intergenerational clinical trials—trials that could result in costly mistakes.

Perhaps the most telling general argument against therapeutic germline interventions lies in the fact that it will prove impossible to limit the technology to genuinely clinical applications. There will be constant pressure to utilize germline modifications for purposes other than the treatment of recognized genetic disease. The fact that the line between the two is blurred should not lead us to conclude that it is insignificant. By contrast, it draws attention to the challenges that would be faced if society decided to employ germline interventions for any purposes, including therapeutic ones.

The broad context of all such interventions—for therapy or enhancement—is that of eugenics. Techniques such as those necessary for germline interventions would concentrate power in the hands of a few people, and such power would either likely or necessarily be corrupting. Just a few individuals could set the course for the genetic modification of many people, and the values held by those people in making the modifications would be passed down to all subsequent generations. Whether such interventions are well intentioned or malevolent, their intent is to recreate the species by design according to a template that is constructed by certain persons, or the fashionable assumptions of one particular generation. The contrasting argument is that every human being has the right not to have been so designed.

These are all general arguments with deep roots in ethical reflection. However, there are other, specifically theological reasons that undercut the case for the practice of germline genetic engineering, beginning with the doctrine of creation and the implications of the incarnation of Jesus Christ, flowing through a Christian understanding of creation, and culminating in a specific delineation of Christian vocation in a fallen world.

## Bioethics and Christian Anthropology

While the questions addressed in bioethics in general seem diverse, they may be reduced to one, for at each stage the contemporary discussion of bioethics is in fact the discussion of how we should treat human beings in relation to two contexts: developments in medicine, the life sciences, and related technologies; and human nature itself. That is, bioethics stands at the interface of medically related technologies with their manipulative capacities, and human nature. The bioethics agenda therefore reduces to an exploration of what human nature means since that meaning will necessarily determine how it is considered that human beings should be treated. While that is generally true, and shifts in the approach to issues in bioethics reflect shifts in general cultural assumptions about human being, it is preeminently true in the context of the Christian religion because Christianity claims at root to teach a fundamental understanding of human being, *coram Deo*.

Christian anthropology is anchored in the two foundational doctrines: those of creation and incarnation. Human beings are created in the image of God. Jesus Christ, the Son of God, took human form in the incarnation. There is mystery, but also logic, in incarnation. If it is granted that human nature is made in the divine image, then for the divine to take human nature to himself is not irrational. While the full meaning of the *imago Dei* in Genesis 1 is not spelled out, it is illustrated almost beyond belief in the story that after many foreshadowings (all the way from the *protevangelium* of Genetic 3 onward) has its beginnings in the Annunciation, for the second person of the triune godhead takes human flesh; the image of the invisible God takes the form of his human image. Human nature is thus premised of its own creator. While the tendency of many Christians to a docetic reading of the incarnation remains strong (that is, the view that the "human nature" of Christ was merely an illusion), the vigor of incarnational christology as the substrate of Christian thinking cannot be better demonstrated than in its radical implications for the bioethics agenda.

To the givenness of created human nature is added, as it were, its takenness, for God having bestowed it has now adopted it as his own. While it is common for Christians to assume that the human nature of Christ ended with his human life in Palestine, this is by no means the case. The classical Christian belief is that having been raised from the dead and ascended into heaven, he "sat down" at the right hand of God the Father. This distinctive doctrine, known as the "session" (sitting) of Christ, serves as a lynchpin of christology since it establishes the continuity of our Lord's human nature in time and eternity. In whatever degree of mystery it is couched, by affirming that our great high priest who has passed into the heavens retains his human nature, it underscores both our access to God through the mediation of one who is still one of us and the defining character of that human nature for our own, since we are now twice declared to be, at least in the analogy of his being and ours, one of his. Thus is asserted the ontological distinctiveness of human being as something other than a mere accident of space and time. Human nature has been chosen by God to be ours, and also to be his own. As Charles Hodge, renowned theologian at Princeton University during the nineteenth century and preeminent American thinker in the Calvinist or Reformed tradition, wrote: "[T]his supreme ruler of the universe is a perfect man as well as a perfect God."[12]

The determinative significance of this fact for every proposal for human enhancement is plain. The analogy of human nature is anchored not simply in its createdness but in its being taken by God for his own. In the Christian eschatology, this same Jesus who is taken up into heaven is the Jesus who will return, and he will return in the glorified but still human form of a Palestinian Jew of the first century A.D.

From this perspective, it is plain that all efforts at the enhancement of human nature—with enhancement defined in terms of a break with the human analogy—are theologically excluded since they have the effect of reshaping that human nature that is both God given and God taken. The exemplar of *Homo sapiens* is the glorified Jesus Christ, and he it is who will return to be our judge. His bearing our humanness sets the standard of all excellences in time and space, and while our humanity (and, as has been argued by an important minority in the Christian tradition, also his) is fallen and subject therefore to both sin and its consequences, every effort at the enhancement of our human nature as such is doomed to failure. The only way for humans to rise above the givenness of their human station must be illusory; the way up leads as it were, only down. When humankind first sought to rise above it edenic station and claim equality with God, the result was the Fall. To make the same point in terms of the biotechnological metaphor: Huxley's *Brave New World* is the great dystopia of our age. Its efforts, like those of all utopians, may come from the best and most optimistic of motives, but its fate is to fail. While there are many paths to the amelioration of our fallen state, they do not lie in recreation at our own hand.

This principle is starkly illustrated in the two great technological achievements of the ancient world: the ark of Noah, and the tower of Babel. The God who provides of his own grace the beneficent technology of the ark to save humankind from his own wrath reveals in his response to the other great technological project of the biblical world, the tower of Babel, the shadow side of all merely human *techne*. At Babel, the human race asserts its independence of God and its intent to use technology to make a name for itself, technology here serving as the symbol of all autonomous human endeavor.

In striking contrast to Cain and Abel and the causes of the Flood, the focus here is not on violence, but on something very different. Indeed, by contrast with this story, Cain's homicide seems tawdry. This is sin of another order. There is no violence here, no illicit sexuality, no worship of false gods. There is simply a building project, one of two great technology projects of the ancient world that are described for us in the book of Genesis. It may perplex us that it could have had the significance that is given to it in Genesis, both by its builders and by God. It is actually helpful if it does, because in that case we are prepared for the problem we have in assessing the technologies of our own day and their often hidden significance—hidden from those who develop them, hidden also from Christians, not least, as they observe and seek to make sense of what is happening. But not hidden from God.

Is Genesis therefore telling us that technology as such is bad? That to seek to extend our powers by simple or complex means is an affront to God and denies our dependence on him? The contrast in Genesis itself is informative. There is not one great technology project in the book of Genesis, there are two. The Flood tells the story of God's judgment on the spreading violence that stemmed from the killing of Abel by his brother. It dominates the early part of the book of Genesis, spanning three chapters and leading into the covenant with Noah. From another perspective the story of the Flood is actually the story of the ark because the ark is the other great technology project of the ancient world. Its building is described in far more detail than the building of the tower.

As we seek to gain a biblical perspective on the explosive power of technology to change our culture and even ourselves, this is where we need to start. One project symbolized humankind's rebellion against God and was answered with massive force by God. The other was a gift of God to humankind, to rescue the human race from the consequences of their own sin. Babel symbolized worldwide rebellion against God and humankind's sinful determination to use technology to go its own way. The ark was technology given by God to preserve the life that he had given.

The contrast could not be stronger. The Babel principle is that of technology out of control, intended to enable humans to have power and achievement entirely apart from God. It led to the scattering of the nations and the curse of enmity and division that has plagued the world ever since. Emerging technologies such as germline engineering claim powers over our own species that will enable us to recreate ourselves in our own image.

The Babel principle returns when humankind decides to exploit the God-given gifts of skill and strength and the plentiful resources of God's world to achieve power through technology. Every previous opportunity that humankind has faced to employ our skills to challenge the authority of God—from Babel on—has only helped to pave the way for the greatest challenge. That challenge comes not in the form of killing and destruction—from the crime of Cain to the widescale violence that brought the Flood on the earth of Noah's day and has deluged the earth in our own time with the atrocities of Auschwitz and Rwanda and Bosnia. It is rather the subtler and most sinister challenge of all—the threat to seize the place of God the creator in designing and redesigning human nature itself. That is the final embodiment of the sinful challenge to God: to use "our" technology to displace him; to make a name for ourselves in this, his world; to let loose the Babel principle in the technology of today.

The reflexive use of these technologies therefore represents the first decisive step across the line that separates the kind of beings we are from the kind of things we make. Thus *Homo sapiens*, who has always been *Homo faber*, humans as makers, by turning our making on ourselves in the sublimest of ironies in a single fateful act both elevates the human self to the role of creator and degrades that same self to the status of a manufacture.

This act is stupefying in its scope. Humankind simultaneously claims the role of God while being reduced to playing the part of an artifact, the dust of the earth out of which we were made and to which we foolish creatures choose to return ourselves; to become commodity rather than creation, made rather than begotten.

From a theological perspective, the significance of both sides of the coin is plain. In our attempt to serve as our own creator we are revealed as usurpers, capable only of manufacture. That Faustian bargain is the only one on offer. The task of creator is personal to God, and his election of the interpersonal mystery of human sexuality as the context for procreation preserves his creatorhood absolutely. The most that his human creatures can do is, as we say, ape his role, parody it, and reduce it to the mechanistic and industrial processes at which we are so good and for which indeed, among other things, we were created. The ambiguity of the designed human, as both creature and product, *Homo sapiens* hijacked by *Homo faber*, moves us decisively toward what the posthumanists call the singularity—that state they envisage in which the distinction between human being and manufactured being is over and a seamless dress weaves together our humankind and what we have made. It anticipates the union of "mecha" and "orga."[13]

In his jeremiad, "Why the Future Doesn't Need Us,"[14] Bill Joy, cofounder of Sun Microsystems, claims that genetics, robotics, and nanotechnology are the three great threats to the human race in the twenty-first century. Through some mixture of accident and intent they are likely to destroy the human species, or supplant it, through some biological or mechanical meltdown or through the triumph of machine intelligence. One does not need to buy the whole thesis to acclaim his comprehensive framing of the issues. At the heart of this secular analysis lies what Christians recognize as a single theological issue: the threat to human nature that is posed by fallen human creativity; the dominion mandate from Genesis 1 to subdue the earth divorced from its biblical context—human dignity made in the image of God.

With germline modification we therefore cross the Rubicon. We venture for the first time into the reengineering of our own kind. Until now, our imaginative depravity had to be content with new forms of killing, the legacy of Cain and Abel. We confront now a new kind of sin, a fresh fulfillment of our conflicted fallen nature, the descendant of the tower of Babel.

In an essentially spurious attempt to find theological justification for such an approach, some writers have coined the term "co-creator" in an effort to acknowledge the enormity of the human claim to (re)shape human nature, and yet dignify it with a designation that, as it were, seeks to bridge the chasm between what is proper to God and what is proper to humankind, blurring the divide between the creator's prerogatives and the dominion that his human creatures are called to exercise. It is, we might venture to say, emblematic of the ambiguities into which fallen human nature has entered.

The concept of human beings existing as created beings whose purpose is to continue to create in partnership with God is one that has become popular in recent years and is preeminently argued for by Lutheran theologian Philip Hefner. In this understanding of theological anthropology, human beings were created by God in the image of God, and that image consists at least in part in the capability of humanity to imagine, to bring into existence that which was not—to create. Humans are "creatures of nature who themselves intentionally enter into the process of creating nature in startling ways."[15] Human beings are capable of creating technologically and thereby recreating or altering the environment of human existence. Hefner links this understanding to the account of the creation of humanity in Genesis 1, in which human beings are given dominion over animals and the natural world. He views this role of human beings as created co-creators as one of both freedom and responsibility, bound up with genetics and culture.[16] He ties this to both scientific and religious aspects of human life in seeking to bring theological reflections to bear on scientific endeavors, such as questions about the meaning and purpose of human life, and guiding responsible human choices for the future of humanity and the natural world.

As created co-creators, endowed with both the freedom to create within the natural world and the responsibility to create wisely and carefully, Hefner considers the technologies created by human beings to be mirrors of ourselves and our desires. This raises serious questions, particularly when human beings try to take on the task of recreating ourselves (and, indeed, our posterity). He sees some of the problems inherent in this explication of the task: "We see in the techno-mirror that although we are busy creating new realities, we do not know why we create or according to what values."[17] This should stand as a caution to those within the theological community who seek to build on this understanding of theological anthropology, since, as it were, it proves too much. Are we free to create as we please? We must remember our responsibility as well as our freedom, our createdness and sinful finitude as well as our dominion. In that light, this coinage is revealed as more of a hindrance than a help. There is no doubt that the human imagination is possessed of great power and that human creativity, for good and for ill, is almost boundless. Yet its context and its control lie in the concept of dominion, that highly specific Genesis concept that spells out the human role within the framework of our stewardship of God's world, a dominion never intended to be one of creative power over other human beings. To move from stewardly dominion as a model to one of co-creation is to invite just that abuse of human freedom that is chronicled in Genesis 3 and what follows. It is a claim, however limited, to equality with God, and it needs to be resisted.

## Enhancement and the Human Analogy

The President's Council on Bioethics' recent report, *Beyond Therapy*,[18] sets out a comprehensive reflection on the move from therapy to enhancement in the prospect (and, to some degree, the present capacity) of biotechnology. It begins with the "therapy–enhancement" distinction, but recognizes that it is not ultimately adequate to the task. Setting out the key problem, the council writes:

> We want better babies—but not by turning procreation into manufacture or by altering their brains to give them an edge over their peers. We want to perform better in the activities of life—but not by becoming mere creatures of our chemists or by turning ourselves into tools designed to win or achieve in inhuman ways. We want longer lives—but not at the cost of living carelessly or shallowly with diminished aspiration for living well, and not by becoming people so obsessed with our own longevity that we care little about the next generations. We want to be happy—but not because of a drug that gives us happy feelings without the real loves, attachments, and achievements that are essential for true human flourishing.[19]

There is, of course, an intentional ambivalence in each of these statements since something in each of us would seek the end without regard for the means; and yet, in most of us there is a stronger intuition that declares the means to be central to the proper attainment of the end. We reflect on the stories of the heroic and the defiant that we want our children to read, on the lives of courage and accomplishment that we seek for them. We muse on the accolades that we covet for ourselves. We discover that whatever our religious or nonreligious understanding of the world, whichever location we find for ourselves on the cultural spectrum, and however we tend to favor or suspect the latest in technology, there is in most of us a hard core of commonality. We admire striving; we despise those who cheat; we applaud the extraordinary achievements of those who triumph over adverse and desperate circumstances; we seek an understanding of our own lives in heroic terms, as those who might one day be said to have fought the good fight and kept the faith, whatever that faith may have been. We touch bottom in a common acknowledgement of what it means to be human, and for all our diversity we grasp human greatness when we see it. We hold Mother Teresa and Abraham Lincoln among our heroes. We watch Tolkien's extravaganza *The Lord of the Rings*, Mel Gibson's *Braveheart*, and Liam Neeson's *Michael Collins* and remember the admonition attributed to Rabbi Hillel, "In a place where there are no men, strive to be a man."

The council's report continues:

> In enjoying the benefits of biotechnology, we will need to hold fast to an account of the human being, seen not in material or mechanistic or medical terms but in psychic and moral and spiritual ones. As we note in the Conclusion, we need to see the human person in more than therapeutic terms: as a creature "in-between," neither god nor beast, neither dumb body nor disembodied soul, but as a puzzling, upward-pointing unity of psyche and soma whose precise limitations are the source of its—our—loftiest aspirations, whose weaknesses are the source of its—our—keenest attachments, and whose natural gifts may be, if we do not squander or destroy them, exactly what we need to flourish and perfect ourselves—as human beings.[20]

The council goes back to Aldous Huxley as their point of reference, with their intuition that the naïve predictions of bliss that will result from an unfettered application of these new technologies will come unstuck in "the humanly diminished world portrayed in Huxley's novel *Brave New World*, whose technologically enhanced inhabitants live cheerfully, without disappointment or regret, 'enjoying' flat, empty lives devoid of love and longing, filled with only trivial pursuits and shallow attachments."[21]

The council assumes that simply to speak in terms of therapy versus enhancement does not seem to work. The line seems too blurred as one person's therapy becomes another's enhancement (growth hormone or, indeed, neuroprostheses, offer nongenetic examples). Yet the line is also fundamental in sketching the point at which the human condition begins to come under threat. One way in which we may articulate the question of human nature without falling into mechanical concepts of where therapy ends and enhancement begins is in terms of analogy. Technological interventions, if they are to sustain and not compromise the human condition, need to retain congruence, as it were, with the human and not trespass upon its analogical integrity. The analogy of human nature offers one means of construing the givenness that we inherit as biological human beings who are members of the species *Homo sapiens*. While a comprehensive definition of what it means to be human escapes us, that does not render us unable to address the question. We may not comprehend, but we may seek to apprehend, the human. While they may not amount to the kind of tight definition that would be required in the preamble to legislation, our stories of heroism and tragedy—from Hebrews 11 to Shakespeare to the news reports of *New York Times* and the all-too-human quirkiness of the cartoons of the *New Yorker*—afford us powerful defining marks for our common humanity made in the image of God. This central recognition on our part, bounded on one side by our shared notions of heroism and achievement and on the other by the ambiguities that subsist in the metaphor of such subhuman exigents as steroids in sports or Viagra for sexual performance helps frame the human question.

## Conclusion: Technology, Anthropology, and Calling

According to Matthew 25:34ff, Jesus describes the end of history this way: "Then the king will say to those at his right hand, 'Come, you that are blessed by my Father, inherit the kingdom prepared for you from the foundation of the world; for I was hungry and you gave me food, I was thirsty and you gave me something to drink, I was a stranger and you welcomed me, I was naked and you gave me clothing, I was sick and you took care of me, I was in prison and you visited me'."

Those who suggest that the fact that the line between therapy and enhancement is blurred would argue the impossibility of denying enhancement options and in the process weaken the case for the prohibition of germline interventions as such. In fact, it does no such thing. It draws attention to the growing problematic nature of interventions (hormonal and surgical, as well as genetic), some current and others in prospect, by which we effect changes in human morphology and capacities, as well as the far greater challenge posed by technologies that will facilitate inheritable, multigenerational change, in what we might describe as a self-directed Lamarckian vision for the human future. By acknowledging that the line between therapy and enhancement is at many points ambiguous, we do not abandon the need to make this distinction or the extreme importance of our seeking to do so. As it happens, the failure of the therapy–enhancement distinction has two effects: It refocuses us on the need for other, less mechanical, models of human flourishing such as the human analogy, and it gives added weight to other considerations that would militate against any germline interventions, for any cause.

General concerns about exercising design power over future generations are restated dramatically in the context of Christian theology, specifically the Christian view of human nature and of our Lord's having taken that nature to himself. It is supremely in the incarnation that we see what it means to be human, and the human nature of Jesus has been taken into the very godhead, unchanging until the eschatological consummation but ready to return at the appointed time with glory.

As followers of Jesus Christ, we have practical ethical callings that must frame our consideration of germline interventions and our response to those in whom such modifications may have been made. Our understanding of vocation includes caring for the least among us, caring for the poor, and loving one another. Our responsibilities are not limited to these callings, but they may point us toward what courses of action we might take in working within the church to respond to pressures to make germline modifications.

The first calling is to care for the least among us. Throughout Scripture, the people of God are called upon to welcome the stranger, to care for the widow and the orphan, and to give a voice to the voiceless and powerless. In the parable in Matthew 25, Jesus specifically mentions that those who inherit the kingdom are those who cared "for the least of these who are members of my family" and that they also cared for him.[22] He names those who are hungry, thirsty, naked, strangers, the sick, and those in prison. We are to care for those who are in need, including those who have genetic structures we would consider to be defects. As the church we must collectively consider how we use the resources we have, especially within the context of wealthy western churches, in which the way we choose to use these vast resources has implications for the future of humanity. It is our responsibility to use our time and resources to love and care for these persons, doing what we are able to do to serve them in their need, and not to decide who needs to be "fixed" in order to be considered worth living. This calling is one to give appropriate medical care to those in need and not to devalue the person who has a need.

As we welcome the strangers of the next generation, they are our guests and not our creatures. The eugenicism that wormed its way into the soul of the early twentieth century church, especially mainline Protestantism, was born of deep confusion and sin in its vision of the nature of the other, and especially the other who is weak or unattractive. Needless to say, as we welcome the stranger we are not to use him or her as the subject of our genetic experiments, not least in the experimentation that would be required by germline interventions, which by its nature would be irreversible and inheritable. Those who would be the subjects of such experimentation would have no choice in the manipulations done to them in the germ cells of their parents. It is the vulnerability of the not-yet-existent, and it is incumbent on those whose existence is established to care for the generations yet unborn with a vigor born of intergenerational accountability. Sondra Ely Wheeler has stated that "Parenting is the most routine and socially essential form of welcoming the stranger."[23] It is this critical task of caring for children in need, whether our own or those of others, that falls to the Christian community in an age of genetic engineering. While others may assume that children will be products that are designed and specified to certain standards, and may well be rejected if they do not meet them, we welcome strangers and therefore our Lord.

The second calling is that of caring for the poor. Technologies such as germline interventions create a class differential that is not merely monetary. Biologist and futurist Lee Silver has looked ahead to the future of genetic technologies and seen a world in which those who can afford it genetically modify their children to give them perceived advantages, including those not available to any human beings today. However, such modifications will not be available to the vast majority of the human community, of course, who will be economically excluded and left in the genetic dustbin. In this futurist vision, the "gen rich" and those without such modifications could quickly evolve to such a point as to be unable to reproduce with one another, effectively creating separate species within humanity. This genetic apartheid finds no support in the Christian calling to fight the exploitation of the poor.[24] This radical commodification of human life would leave the power and resources to direct human genetic history ever more concentrated in the hands of a few.

The third calling that lies behind all others is the Christian calling to love one another. In these technologies we see the ultimate temptation to obtain power over others for our benefit and for what we take to be theirs. We are called to avoid the generation of genetic castes, with great divides of genetic wealth and poverty among human beings; and we must also avoid the tyranny of a single generation over all subsequent generations.[25] The act of parenting—begetting and conceiving—must be distinguished from the act of designing what human beings will be down the generations. We have the power to instill our flawed and culturally conditioned genetic preferences and values in the next generation and so exercise a tyranny over all future generations of human beings, who would become the genetic products of our own devising.

Aside from prudential and general ethical considerations, the prospect of intervention in the human germline raises in the sharpest terms the proper place of humankind in the divine economy, and the place of Christian anthropology in setting limits to our use of technology on our own selves. The moral basis of disease and death within biblical theology, and the concomitant redemptive underpinnings of their final overthrow at the *eschaton*, set the framework within which medicine and the healing arts are to be engaged and within which the proper exercise of human *techne* is to be understood. Just as our dominion of the nonhuman created order is to be exercised as a stewardship, so our stewardship, as it were, of our own selves is to be exercised as something other than a dominion. It is for this reason that the coinage by some of co-creator language, while at one level merely the use of metaphor, is at another level a fundamental assault on the distinctions set out in the opening chapters of Holy Scripture, and an invitation to upend the created order in which while God's image-bearing human creatures occupy the highest place in the firmament, they are separated by an unbridgeable gulf from the Creator himself. Indeed, we could argue that the entire discussion of this use of technology could be framed in terms of the employment of co-creator language as either metaphor or hubris because it spans the deep divide between the order of things set out in Genesis 1 and 2 and their radical overthrow in Genesis 3.

The incarnation of Jesus Christ, here as elsewhere, trumps every other consideration. The taking to himself of the flesh and blood of a Palestinian Jew of the first century A.D. underscores the significance of human being in the most emphatic terms. The sanctity of *Homo sapiens* is established as our species and also his.

## Notes

1. For example, see Sharon Begley, "Why gene therapy still hasn't produced forecast breakthroughs," *Wall Street Journal*, February 21, 2005.

2. For example, see Gina Kolata, "Panel to advise testing babies for 29 diseases," *New York Times*, February 21, 2005.

3. Mihail C. Roco and William Sims Bainbridge, eds., *Converging Technologies for Improving Human Performance* (Washington, DC: National Science Foundation, 2002).

4. The following is a case in point: In the fall of 2005 the Genetics and Public Policy Center at Johns Hopkins University hosted an invitational conference on germline interventions, bringing together more than seventy leading participants in the discussion as part of a national consultation exercise. The platform of presenters made no reference whatever either to other technologies in general and their capacity to introduce species-altering effects, or specifically to the NSF's converging technologies process, of which biotechnology is one component. It was left to one of us (NC) to raise the question from the floor. The panelists did not respond.

5. For a fuller discussion, see Nigel M. de S. Cameron, *The New Medicine: Life and Death after Hippocrates* (Wheaton, IL.: Crossway Books, 1991, reprint Chicago: Bioethics Press, 2001), pp. 171–182.

6. Ronald S. Cole-Turner, "Is genetic engineering co-creation?", *Theology Today* 44:3 (1987): 344–345. Here Cole-Turner cautiously argues that genetic engineering and specifically germline interventions, may be seen in this light. He uses the term "co-creators" from a 1982 report on genetic engineering published by the National Council of Churches of Christ, USA.

7. Cole-Turner, "Is genetic engineering co-creation", pp. 344–345.

8. C. S. Lewis, *The Abolition of Man* (New York: Collier Books, 1962), p. 65. The original essay, "The Abolition of Man," first appeared in 1943 as "Reflections on Education with Special Reference to the Teaching of English in the Upper Forms of Schools."

9. Lewis, "Abolition of Man," p. 69.

10. Lewis, "Abolition of Man," pp. 70, 72, 77.

11. LeRoy Walters and Julie Gage Palmer, *The Ethics of Human Gene Therapy* (New York: Oxford University Press, 1997), p. 80. Walters and Palmer offer a definitive and fair summary of the arguments on all sides.

12. Charles Hodge, *Systematic Theology* (New York: Charles Scribner's Sons, 1885), vol. 2, p. 637.

13. These useful terms are from Spielberg's movie *AI Artificial Intelligence*, generally memorable only for its special effects.

14. Bill Joy, "Why the future doesn't need us, *Wired* 8.04 (April 2000). Available at http://www.wired.com/wired/archive/8.04/joy.html (accessed May 7, 2007).

15. Philip Hefner, "Cloning as Quintessential Human Act." Chicago Center for Religion and Science, 1997.

16. Philip Hefner, *The Human Factor: Evolution, Culture, and Religion* (Minneapolis: Fortress Press, 1993).

17. Philip Hefner, "Technology and human becoming," *Zygon* 37:3 (2002): 660.

18. President's Council on Bioethics, *Beyond Therapy: Biotechnology and the Pursuit of Happiness* (Washington, DC: U.S. Government Printing Office, 2003).

19. President's Council, *Beyond Therapy*, p. xvii.

20. President's Council, *Beyond Therapy*, p. xvii.

21. President's Council, *Beyond Therapy*, p. 7.

22. Matthew 25:40, New Revised Standard Version.

23. Sondra Ely Wheeler, "Making babies? Genetic engineering and the character of parenthood," *Sojourners* (May/June 1999), p. 14.

24. This is related to the argument in favor of skepticism regarding power relations that involve genetic technologies made by Jackie Leach Scully in "Women, theology, and the Human Genome Project," *Feminist Theology* 17 (1998): 69.

25. This notion of biological tyranny, outlined as we have seen by C. S. Lewis, is explored more fully by Bill McKibben in *Enough: Staying Human in an Engineered Age* (New York: Times Books, 2003).

# 6

# Human Germline Therapy: Proper Human Responsibility or Playing God?

James J. Walter

The two concerns of this chapter are primarily theological in nature and scope, although both entail ethical issues. First, I want to show that the moral judgments that religious believers[1] arrive at on the topic of human germline therapy are informed by and at least partially dependent on specifically theological beliefs about God and the nature and destiny of humanity.[2] The second theological concern is to decide whether we are really playing God by manipulating our genetic code in germline therapy or whether such interventions are only another way of properly exercising human responsibility.

James Watson, the first director of the U.S. Human Genome Project, recognized from the beginning of this scientific venture that there were many issues of a nonscientific nature connected with the project. He urged that at least 3 percent of the genome funds ($90 million) be spent on examining these important issues. He succeeded in his efforts, and so the Joint Working Group on the Ethical, Legal and Social Issues Relative to Mapping and Sequencing the Human Genome (ELSI) was formed and began its work in September 1989.[3] Watson was indeed correct about the relevance of the ethical issues connected to this initiative. The scientific breakthroughs that are being made today because of this research, and those that will be made in the future, present us with extraordinarily important and far-reaching moral questions. Before addressing some of these questions, I will quickly review the various types of genetic manipulation that will likely be possible as a result of the mapping and sequencing of the human genome that was completed in April 2003.

Medical scientists could conceivably develop four different types of genetic manipulation from the results produced in the Human Genome

Project.[4] The first two types are therapeutic in nature because their intent is to prevent or to correct some genetic defect that causes disease. The other two types are not therapies at all. Rather, they are concerned with improving either various genetic aspects of the patient him or herself (somatic cell) or with permanently enhancing or engineering the genetic endowment of the patient's children (germline).[5]

The first kind of genetic manipulation is somatic cell therapy in which a genetic defect in a body cell of a patient could be corrected by using various enzymes (restriction enzymes and ligase) and retroviruses to splice out the defect and to splice in a healthy gene. Medical scientists have already used a variation of this technique to help children in France who suffered from X-linked severe combined immune deficiency. Estimates are that there are between three and four thousand different genetic diseases,[6] and these diseases afflict approximately 2 percent of all live births.[7] It is clear that the ability to correct these defects would benefit many patients and save billions of dollars in health care costs over the lifetime of these patients. Second, and this is the only subject of my analysis, there is germline gene therapy in which either a genetic defect in the reproductive cells—egg or sperm cells—of a patient would be repaired or a genetic defect in a fertilized ovum would be corrected in vitro before it is transferred to its mother's womb.[8] In either case the patient's future children would be made free of the defect by permanently altering their genetic code.[9]

Next are the two kinds of nontherapeutic or enhancement genetic manipulation. The first is enhancement somatic engineering. In this type, a particular gene could be inserted to improve a specific trait, for example, either by adding a growth hormone to increase the height of a patient or by genetically enhancing a worker's resistance to industrial toxins. Second, there is germline genetic engineering in which existing genes would be altered or new ones inserted into either germ cells or a fertilized ovum so that these genes would then be permanently passed on to improve or enhance the patient's offspring. In this last form of genetic manipulation, parents could design their children according to their own desires. Although modification of the germline for enhancement purposes is an extremely important topic, my analysis here does not focus on this technology.

## Moral Dimensions

Before addressing the specifically theological issues that serve as the context for moral decision making on germline therapy, it might be helpful to review some of the moral dimensions of this topic. First I analyze five of the moral arguments against and five of the moral arguments for intervening in the germline for therapeutic purposes. Second, because I write as a Roman Catholic theologian, I discuss from this religious perspective some of the moral themes connected to the topic.

Most authors[10] and most national and international commissions or councils of a civil[11] or religious[12] nature have argued morally against any form of enhancement genetic engineering (somatic or germline). In addition, most of the same authors and commissions or councils have argued in favor of pursuing research and implementation of somatic cell therapy for serious genetic diseases.[13] Morally the most contentious form of genetic manipulation, then, concerns therapeutic interventions in the germline that aim at preventing or curing a genetic defect of either the reproductive cells or of the zygote before transfer to the mother's womb.[14] Clearly, the controversy cuts across several areas beyond ethical considerations. For example, it necessarily involves the medical and scientific fields because it is not clear whether this technique is technologically feasible without doing great harm to either the patient or his or her progeny.[15] It is also a social or public policy issue because we must be concerned with whether we could ever reach a consensus as a society on the implications of such research and medical interventions that would permanently change our genetic code.[16] Finally, as I have suggested earlier, it involves a theological problem of whether we have now entered the realm of playing God by using this technology.

## Moral Arguments for and Against Germline Therapy

Eric Juengst has helpfully summarized the moral arguments for and against germline therapy.[17] There are five arguments against such interventions. First, there is scientific uncertainty and clinical risks involved with these techniques. Germline therapy would involve too many unpredictable, long-term iatrogenic risks and harms to the altered patients and their offspring to be justifiable.[18] Second, there is the inevitable slippery slope to forms of genetic enhancement engineering. Modification of the germline would soon open the door to nontherapeutic experiments to improve our progeny, and so we should never move onto the slope that will eventually lead us to these enhancement techniques. Third, the future generations that would be experimented on are unable to give their informed consent. Thus, such interventions would violate one of the most sacred moral principles in human experimentation, viz., the principle of informed consent. Next, there is the moral issue of allocation of scarce resources. Germline gene therapy will never be cost effective enough to justify the expense of these techniques in the face of alternative approaches to the genetic problems, e.g., somatic cell therapy. Finally, there is the issue of the integrity of genetic patrimony. All germline gene therapy would violate the moral rights of subsequent generations to inherit a genetic endowment that has not been intentionally altered.

There are also five arguments in favor of such therapeutic interventions. First, there is the issue of medical utility in which such techniques would offer a true cure for many genetic diseases. Therapeutic interventions at any level above the causal gene would only be palliative or symptomatic. Second, this form of genetic intervention may be the only effective way of medically addressing some genetic diseases, and thus it is an argument for medical necessity. Next, there is the argument about prophylactic efficiency. By preventing the passing on of disease-causing genes, germline therapy would eliminate the need to perform costly, risky somatic cell therapy in multiple generations. Fourth, there is the moral need to have respect for parental autonomy. Medicine should accept, and respond to, the reproductive health needs of prospective parents, including any requests for germline therapy. Finally, there is the moral argument to respect scientific freedom aimed at developing such therapeutic techniques, as long as these techniques are pursued within the boundaries of acceptable research on human subjects.

## Moral Themes from the Roman Catholic Tradition

There are six central themes that inform the moral reasoning on genetic manipulation from the official Roman Catholic perspective, i.e., from the magisterial teachings of recent popes, bishops, and the Second Vatican

Council. In general, I find most official statements since Vatican II (1965) to be quite hopeful and favorable toward genetic science with respect to the issue of manipulating the human genome as long as certain moral boundaries are respected.

First, we are permitted to pursue various genetic manipulations as long as we respect the natural law, i.e., the moral law that is inscribed in the nature of humans and their moral acts. In the Catholic tradition, the order of nature grounds human morality, and this morality is not only objective but also in principle capable of being known by all people of goodwill. As Cardinal Karol Wojtyla (Pope John Paul II) claimed in his book *Love and Responsibility*, a rational acceptance of the order of nature is at the same time a recognition of the rights of the Creator.[19] Concretely, the natural law requires that we respect the dignity of each human being, and thus the natural law would prohibit treating humans and embryos from the moment of conception as a means to some other end. Second, the official teachings from the Roman Catholic Church express a strong ethic of stewardship. This ethic points to two things: We have a God-given responsibility for and toward all creation, including our bodies and we are not the owners of our own bodies but only stewards over them, so we are not free to manipulate our genetic heritage (or nature) at will. Third, the human body is not independent of the spirit. Concretely this means that we cannot expect to alter our genes without also altering the body's relation to our spiritual natures, i.e., who we are as a body-soul composite.[20] Fourth, genetic experimentation on human subjects, including embryos from the moment of conception, is permissible as long as "it tends to real promotion of the personal well-being of humans, without harming human integrity or worsening human life."[21] Informed consent from the one experimented on or from a legitimate surrogate is absolutely required for such experimentation.

Some further clarification on this last theme is important in order that there not be any misunderstandings about the official Catholic tradition. There are a number of constraints that would be placed on any attempt to modify the germline of embryos during the period of experimentation. For example, any experimentation on embryos to achieve a genetic modification that would involve their destruction or harm would be prohibited. Furthermore, one must make a clear distinction between the goal of achieving germline modification, which could be in principle approved of, and the means or methods used to achieve this goal, such as direct manipulation of the ex vivo embryo, which might be de facto prohibited in an experiment because it would cause serious harm to the embryo. Thus, the International Theological Commission the Catholic Church has made the following point.

> Germ line genetic engineering with a therapeutic goal in man would in itself be acceptable were it not for the fact that is it is hard to imagine how this could be achieved without disproportionate risks, especially in the first experimental stage, such as the huge loss of embryos and the incidence of mishaps, and without the use of reproductive techniques. A possible alternative would be the use of gene therapy in the stem cells that produce a man's sperm, whereby he can beget healthy offspring with his own seed by means of the conjugal act.[22]

The fifth theme involves the fundamental relationship between scientific research and the common good of society. This clearly indicates that all such efforts to manipulate the human genome not only involve ethical issues but also have public policy implications. Finally, not every scientific advance necessarily constitutes real human progress. Genetic manipulation to influence inheritance that is not therapeutic but is aimed at producing human beings selected according to sex or other predetermined qualities (eugenics or enhancement) is judged contrary to the personal dignity of the person and consequently contrary to the natural law.

## Theological Reflections

Two decades before the Human Genome Project officially began, the theologian Paul Ramsey warned us about the possible developments in genetic engineering. He claimed with great confidence that "Men ought not to play God before they learn to be men, and after they have learned to be men they will not play God."[23] For Ramsey, to play God certainly meant to convey a negative moral connotation, and his theological statement was aimed at limiting human efforts in the entire arena of genetic manipulation. For others, to play God means to appropriate for ourselves various functions and tasks that properly belong only to the divine.[24] For some this phrase might mean a reaction to the realization that humans are now on the threshold of understanding how the very building blocks of life work. Such understanding would indeed be awesome and thus could justify the description of being God-like. The phrase could also mean either changing, and thus violating, God's created order by using this new technology or it could mean imitating the Creator by fabricating new life forms through germline engineering.[25] Finally, the phrase is sometimes construed within the theological context of usurping the rights of God over creation, and thus to play God is to act from a lack of a right (*ex defectu juris in agente*) in a certain area of life.[26] In this last sense the phrase connotes affective and attitudinal responses of caution and restraint with respect to God's sovereignty over all creation.[27]

Considering these various meanings of playing God, the theological question that I would like to pursue is the following: Is performing germline therapy on humans contrary to God's intentions and purposes and therefore an act of usurping God's rights over creation? An affirmative answer to this theological question would almost inevitably translate into an absolute moral prohibition against all germline therapeutic interventions. On the other hand, a negative response might morally permit these genetic interventions, but it need not result in such a moral judgment. A negative answer might only be a judgment that such therapeutic actions would not be prohibited on strictly theological grounds alone. Therefore, an action could be judged in general as permissible on strictly theological grounds, i.e., it is not contrary to God's intentions and purposes, but currently is not permissible on moral grounds, i.e., owing to current scientific or technical limitations the action might violate the moral principle of nonmaleficence because it harms either the patients themselves or their future progeny.

Like most moral judgments, an answer to the theological question posed here would have to be decided for believers within the broader context of a religious interpretation of experience. Christians, at least, have regularly thematized their experiences of the divine and expressed them through certain doctrinal themes in terms of the Creation, the Fall, incarnation, redemption, and eschatology.[28] These doctrinal expressions themselves have been based on certain models of God and of how the divine relates to and acts in nature and history. In addition, these themes, which have conveyed the Christian interpretations of God, have also served as anthropological frameworks for understanding our moral obligations for both the present and the future. I will use the framework of these fivefold Christian mysteries[29] to show how moral judgments on germline therapy rely on and are authorized by certain theological beliefs and interpretations.[30] However, there is only space to develop the essential lineaments of the various positions under each Christian theme. I formulate my own position on the question of whether or not these therapeutic interventions are contrary to God's purposes by stating under each theological theme the position I adopt.

**Creation**

The doctrine of creation is actually a complex set of interpretations of who God is and how the divine directs human history and acts within it (divine Providence). These theological interpretations have anthropological counterparts that attempt to understand both how we as created beings stand in the image of God (*imago Dei*) and how we are to evaluate the significance of physical nature and our bodily existence.

Two different theological models of God, creation and divine Providence, have been historically used in the great Christian tradition. Currently, Christians have used both models as a theological context in arguing morally for or against germline interventions to cure serious diseases.

In one perspective God is viewed as the creator of both the material universe and humanity and the one who has placed universal, fixed laws in the very fabric of creation. This view of creation obviously favors Parmenides' interpretation of reality as fixed and static,[31] and it assumes that God's purposes for humanity, which are forever unchangeable, can be known by reflecting on the universal laws governing nature and humanity. As sovereign ruler over the created order, God directs the future through divine Providence. As lord of life and death, God possesses certain rights over creation, which in some cases have not been delegated to humans for their exercise.[32] When humans take it upon themselves to exercise God's rights, for example, those divine rights to decide the future or to change the universal laws that govern biological nature, they usurp divine authority and thus they act contrary to God's purposes in creation. If one adopted the theological positions held in this model, one would likely judge as human arrogance the attempt to alter the genetic structure of the human genome, even to cure a serious disease. This assessment is confirmed in a Time-CNN poll on people's reaction to genetic research. Not only were many respondents ambivalent about genetic research, but a substantial majority of the respondents (58 percent) thought that altering human genes in any way was against the will of God.[33]

In the second theological model, which I adopt, God is not interpreted as the one who has created both physical nature and humanity in their complete and final forms. Rather, the divine continues to create in history (*creatio continua*). Consequently, God is not understood as having placed universal, fixed laws in the fabric of creation, and so the divine purposes are not as readily discernible as in the first model. God's actions both in creation and in history continue to influence the world process, which is open to new possibilities and even spontaneity. Divine providence is understood as God providing ordered potentialities for specific occasions and responding creatively and in new ways to the continually changing needs of history.[34] Although there is some stable order in the universe, like Heraclitus' view of all reality, creation is not finished, and history is indeterminate. Because creation was not made perfect from the beginning, one can discern certain elements in the created order, such as genetic diseases, that are disordered. Because these disordered aspects of creation cause great human suffering, they are judged to be contrary to God's final purposes and so can be corrected by human intervention.[35]

As an anthropological counterpart to their interpretations of the divine, Christians, like Jews, have consistently understood all humanity to be created in the image and likeness of God (Genesis 1:26–27). However, the great Christian tradition has used at least two different interpretations of how humans stand in that image,[36] and these diverse models almost inevitably lead to different moral evaluations about therapeutic interventions in the human genome.

The first interpretation defines humanity as a steward over creation. Our moral responsibility, then, is primarily to protect and to conserve what the divine has created and ordered. Stewardship is exercised by respecting the limits placed by God on the orders of biological nature and society.[37] It is easy to see how this model is consistent with the understanding of God as the creator who has placed universal, fixed laws in the very fabric of creation. If we are only stewards over both creation and our own genetic heritage, then our moral responsibilities do not include the alteration of what the divine has created and ordered. Our principal moral duties are to remain faithful to God's original creative will and to respect the laws that are both inherent in creation and function as limits to human intervention.

The second interpretation of the *imago Dei* defines humans as co-creators[38] or participants[39] with God in the continual unfolding of the processes and patterns of creation. As created co-creators[40] we are both utterly dependent on God for our very existence and simultaneously responsible for creating the course of human history. Although we are not God's equals in the act of creating, we do play a significant role in bringing creation and history to their completion.[41] Because I adopt this position, I would argue that part of our responsibility in bringing creation to its completion might even include permanently overcoming the defects in biological nature that remain contrary to God's purposes.

A Christian interpretation of the significance and value of both physical nature and our bodily existence also plays an important role in arriving at moral judgments about genetic therapy. There are several different models of material nature that can shape one's moral position on genetic manipulation. Each model attempts not only to interpret the nature of all material reality but also to understand the extent to which we can use human freedom to change our biological processes for therapeutic purposes.

Daniel Callahan has argued that one of the most influential models of nature that operates in contemporary society is the power-plasticity model. In this view, material nature possesses no inherent value and it is viewed as independent of and even alien to humans and their purposes. All material reality is simply plastic to be used, dominated, and ultimately shaped by human freedom.[42] Thus the fundamental purpose of the entire physical universe, including human biological nature, is to serve human purposes. What is truly human and valuable are self-mastery, self-development, and self-expression through the exercise of freedom. The body is subordinated to the spiritual aspect of humanity, and humans view themselves as possessing an unrestricted right to dominate and shape, not only the body, but also its future genetic heritage.

The view of nature at the opposite extreme is the sacral-symbiotic model in which material nature is viewed as created by God and thus considered as sacred. As created and originally ordered by God, human biological nature is static and normative in this understanding, and the laws inherent in it must be respected. We are not masters over nature, but stewards who must live in harmony and balance with our material nature. Biological nature remains our teacher and shows us how to live within the boundaries established by God at creation. Since physical nature is considered sacrosanct and inviolate, any permanent alteration of the human genetic code, even to cure a serious genetic disease, would probably be morally prohibited.

I am a proponent of the final model, which construes material nature as evolving. Whereas there is some stability to nature and there are some laws that do govern material reality, neither this stability nor these laws are considered absolutely normative in moral judgments. Change and development are considered more normative than other aspects of nature, and history is seen as linear rather than cyclic or episodic.[43] The relation between material nature and human freedom appears as a dialogue that evolves dynamically over time. It is within this dialogue that responsible humans learn how to use material reality as the medium of their own creative self-expression.[44] This model would seem to grant to humans the freedom and responsibility to intervene in our evolving biological nature to correct serious diseases at the germline level. The reason is because such human efforts would not necessarily be judged as usurping God's final prerogatives or purposes in creation.

### The Fall

The Christian tradition has taught that although creation is essentially good, humans have misused their freedom and acted irresponsibly. This teaching, then, refers to a fall that has infected all human history. However, there have been different interpretations about the depth of human depravity and the connection between this fall and all disease, including genetic diseases. In one way, this doctrine functions as a way of assessing the extent to which humans, especially medical scientists, can be morally trusted with the awesome powers to alter the human genome, even for therapeutic reasons.

One view of the human fall, which was adopted by many early Protestant reformers and continues in the thought patterns of some contemporary theologians, is that all aspects of the human person are deeply affected by sinfulness. This interpretation has led some to distrust the claim that humanity will use genetic interventions only for moral ends, e.g., to cure disease. Consequently, proponents of this view regularly try to limit the extension of human control over the genetic heritage of individuals and their progeny for fear that this therapy will inevitably slide down the slope to improper genetic engineering.[45] This view does hold that genetic diseases are contrary to God's original purposes in creation. However, it also tends to connect the origin of these diseases with the misuse of human freedom.

At the opposite end of the spectrum, religious advocates almost entirely forget about the fall of humanity. They look upon the Fall as insignificant and inconsequential, and they consider only the possibilities open to human ingenuity and rational control. Consequently, these proponents regularly support efforts to manipulate the human genome. By downplaying the effects of the Fall on humanity, they extol human freedom and control over physical nature and the future.[46]

I adopt an alternative view to these two extremes. This position, which has been historically consistent with Roman Catholic thought, could be described as a moderately optimistic assessment of the human condition. Though fallen, humanity remains essentially good and can know and do the moral good with the grace of God. Unlike the excessively optimistic view in the second interpretation, adherents of this view recognize that the human capacities to reason and will the moral good continue to be affected by sin. Consequently, they are cautious about putting too much trust in humanity's ability to use modern technology solely for moral ends. However, they do not necessarily view therapeutic interventions in the human germline as violations of God's sovereignty over creation, nor do they judge these efforts to be contrary to divine purposes. In addition, genetic diseases are viewed fundamentally as natural to the unfinished created order and so they do not necessarily originate with human irresponsibility in the Fall.[47] These disorders have always been and continue to be contrary to divine purposes for humans, and so therapeutic manipulations of the human germline to cure them are not in themselves wrong on theological grounds.

**Incarnation**

The fact that God took on human bodily form in the person of Jesus Christ has several implications for the discussion of genetic medicine. First, this doctrine serves as a context both for assessing the relation between body and spirit and for evaluating the significance of the body in moral decision making. In turn, these considerations have an impact on the question of what we judge to be the normatively or uniquely human in moral analysis. Both issues function as presuppositions to moral judgments about the permissibility of germline therapy.

If one separates, or even grossly distinguishes, body and spirit, there is the tendency to view our spiritual part as more important or even as the solely unique characteristic of the human person. In addition, such a view will tend to hold that permanent alterations of the body, e.g., through genetic manipulation of the germline, do not and cannot actually change the fundamental nature of humans. The influential physician-research biochemist W. French Anderson once remarked that he had been worried for years that we might end up altering our very humanness by methods of genetic engineering. However, he has recently decided that Plato was correct to view the soul and the body as two distinct entities.[48] By adopting this Platonic framework Anderson now believes that we cannot alter our fundamental humanness because, as much as we might permanently change our biological genetic code, we cannot change that which is uniquely or normatively human about us, viz., our soul or that which is beyond our "physical hardware."[49] Some contemporary theologians who have addressed this issue of gene therapy have also adopted a similar position on the nature of the human. For example, G. R. Dunstan[50] has argued that only if gene therapy intervened at the level of the cerebral cortex and the central nervous system to alter the capacities of self-consciousness, inquiry, rational ordering and analysis, moral judgment and choice would human nature really be changed.

An opposing view is the position that holds that there is an intimate relation between body and soul. I would argue that we are embodied spirits or ensouled bodies.[51] As Paul Ramsey once phrased it, "We need rather the biblical comprehension that man is as much the body of his soul as he is the soul of his body."[52] Such a view, then, would be far more cautious than the first about making a claim that we cannot permanently alter the nature of humanity through genetic manipulation. The relation of body and spirit is one, but not the only, element of what makes up our fundamental human nature. Thus to alter this relation would imply the possibility of changing our nature in this view. However, since the intent in germline therapy is to prevent or to cure disease and not to enhance or engineer the human subject, I would conclude there is much less risk that we will change this aspect of our human nature, i.e., the body-spirit relation, through this intervention.[53]

**Redemption**

Christians believe not only that we are created yet fallen beings but also that we are redeemed by God through the suffering, death, and resurrection of Jesus Christ. Thus, besides God's creative purposes, the divine also has redeeming purposes for all creation. Christians have sometimes grossly separated the creative and redeeming purposes of God. One way to understand the relation between these divine activities has been to interpret redemption as not only a continuation of creation but also the means by which creation itself is brought to completion by God. This framework raises the question of whether the technology to alter the genetic code for therapeutic purposes can ever be viewed as potential participation in God's redeeming actions toward humanity. Since Christians have interpreted humankind as created in the divine image, it has been possible to view genetic interventions as possible acts of *co-creation* with the divine. However, now the question is whether it is also theologically possible to view our technological activities and interventions as potential participations in or mediations of God's redemptive purposes. Answering this question requires a brief discussion of various theological evaluations of technology in general.

There are several evaluations of modern technology that could serve as the context for our moral judgments about therapeutic techniques to cure serious genetic diseases. First, there is the rather pessimistic view of technology,[54] an example of which the late Jacques Ellul adopted.[55] Its characteristics include a skeptical attitude toward any real benefits from technology and a great sensitivity to the potential evils that will come from its development and use. Technology is viewed as a threat, impersonal, manipulative, and alienating, and thus it does not and cannot possess the inherent potential to share in the divine purposes of redemption, which are personal, salvific, and holistic.

The opposite extreme is an overly optimistic view of technology and its potential achievements.[56] Its hallmarks are a focus on the liberating function of technology through progress and human fulfillment and an emphasis on greater freedom and creative expression. Some, like the Jesuit paleontologist Pierre Teilhard de Chardin,[57] have closely linked technology and spiritual development and thus have viewed technology as clearly possessing the potential to cooperate with God's work.

The final position seeks to steer a middle course between the two extremes of pessimism and optimism. Similar to the first view, its proponents are cautious about and critical of many features of modern technology. However, like the second view, these proponents also offer hope that technology has the potential to be used for humane moral ends, but technology must be redirected in its uses for these ends to be realized. There are two forms of this moderate position currently held by theologians that I would like to analyze quickly. Among other things, these views are distinguished by how they causally connect sin with disease and death. In other words, these positions differ depending on how one interprets St. Paul's passage in the Letter to the Romans (5:12): "It is just like the way in which through one man sin came into the world, *and death followed sin*, and so death spread to all men, because all men sinned."

The first form of this position causally links the introduction of death and all disease, including genetic disease, to the entrance of sin into the world. The role of medicine, then, is to intervene to overcome these effects of sin, and these medical interventions, including those aimed at genetic therapy, are construed as mediations of God's redemptive activity. In this same view, however, all forms of human gene transfer whose primary purpose is to enhance or engineer the human would be at least morally problematic on theological grounds. Why? Because these interventions would not alleviate any condition that can be causally linked to the entrance of sin into the world. Their purpose would be to enhance the patient or his or her progeny, not to overcome the effects of the Fall.[58]

The second form of the moderate position does not causally link sin with disease and thus does not identify disease as such as one of the effects of the Fall. Rather, it understands diseases (and for that matter, death) as the natural results of being part of the material world, where decay and entropy are facts of the created world, although sin may very well adversely affect our experiences of these realities.[59] That does not mean that God wills or permits these ill effects as part of the final divine ordering of the universe; in fact, they are judged to be contrary to God's ultimate purposes. The Protestant theologian Ronald Cole-Turner has adopted a position similar to this one.[60] He has argued that modern technological developments in genetics can have the potential for participating in God's redemptive activities. He has reasoned that when this technology is aimed at preventing or curing serious genetic diseases that are deemed contrary to God's final purposes for humanity because they cause great human suffering, this technology can participate in God's redemptive purposes by making whole and healthy what was once disordered and destructive. Cole-Turner, like the first form of this position, however, does not seem to support human gene transfer whose primary purpose is enhancement, not therapy.

**Eschatology**

The great Christian tradition has affirmed that all creation is called to a future beyond this history, i.e., to an eschatological era as the final end of human history. This future is variously called the "reign of God," the "kingdom of God," or "God's absolute future." However one names it, Christians believe that it is God who inaugurates this future and brings it to final consummation. Interpretations of the relation between our human history and God's eschatological future function as the background context for discerning our moral responsibilities toward the human future. Thus, various eschatological visions will contextualize differently the discernment of our moral obligations to improve our genetic heritage through germline interventions.

Harvey Cox identified three strains of eschatology that traditionally have been used in Judeo-Christian theologies: the apocalyptic, the teleological, and the prophetic.[61] He argued that all three can be found in both ancient religious traditions and modern secularized forms. Each religious strain has a different understanding of God's eschatological future and how God will inaugurate that future. Consequently, each strain will construe quite differently the relation of humanity's historical future to God's absolute future, and each will variously formulate what our moral responsibilities are for making sure human history turns out right.

The apocalyptic eschatology, whose origins are in ancient near-eastern dualism, always judges the present as somehow unsatisfactory. In both its religious and secularized forms, this eschatology evaluates this world and its history negatively and it foresees imminent catastrophe. The religious form of this eschatology always draws a sharp distinction between God's absolute future in the kingdom and the conditions of our human history, and thus it generally argues for a great discontinuity between this world and the next.

On the other hand, the teleological eschatology, whose origins are Greek but which was adopted by Christians, views the future as the "unwinding of a purpose inherent in the universe itself or in its primal stuff, the development of the world toward a fixed end."[62] All creation, then, is moving toward some final end, for example, beatific vision with God. Consequently, there is some continuity between present human history and God's future.

The last interpretation of eschatology, which I adopt, is the prophetic strain. Its origins are Hebrew in nature, and it views the future as the open area of human hope and responsibility. In the Hebrew Scriptures, the prophets did not foretell the future; rather, "they recalled Yahweh's promise as a way of calling the Israelites into moral action in the present."[63] In its biblical form, then, the future is not known in advance, but it is radically open and its actualization lies in the hands of humans, who must take responsibility for it. In its modern secularized form, the prophetic eschatology places great hope in human responsibility for the future, and it views the future with its manifold possibilities as unlocking the determinations of the past.

One of the most notable theologians who had adopted the apocalyptic eschatology and then applied it to issues in genetic research was the late Paul Ramsey. He regularly emphasized the discontinuity between this world and the next, and thus he always urged us to remain faithful to God's future as that is represented in the divine covenant between humanity and God. Ramsey argued that we do not have any moral obligation to safeguard the future of humanity through genetic research because he believed that "religious people have never denied, indeed they affirm, that God means to kill us all in the end, and in the end He is going to succeed."[64] It is this apocalyptic view, which interprets human history as coming to an abrupt end through divine activity, that influenced Ramsey's interpretation of both our general moral responsibilities for the future and his specific moral prohibitions against genetic research that would permanently alter our genetic code. Our primary moral responsibility, in his view, was to remain faithful to what God has given us; it was not to act as if we had the moral responsibility to save our future offspring from genetic disease. If one adopted either a teleological or a prophetic eschatology, one would be inclined to accept certain genetic interventions in the human germline to cure serious disease. Both strains emphasize human responsibility for the future, although each does this differently. Both understand that the future is open and somewhat indeterminate. Consequently, these eschatologies could serve as warrants for morally justifying germline therapy. Neither view would necessarily judge that such interventions would be contrary to God's creative and redemptive purposes, and thus neither would necessarily hold that these techniques would be wrong on theological grounds alone.

## Conclusion

My theological interests in germline therapy have been twofold. I have sought to show how Christian moral decision making on the new genetics is contextualized by specifically theological beliefs. I have also posed the theological question of whether therapeutic interventions in the human germline to prevent or cure serious diseases are acts of playing God, or whether such actions are within the boundaries of authentic human responsibility. An answer to this question, I have argued, must be decided within the broader context of several theological affirmations or doctrinal themes that are interpretations of religious experience.

It is my judgment that significant scientific and technical difficulties remain to be solved in respect to both types of germline therapy and that there continue to be public policy difficulties with these genetic technologies as well. As mentioned earlier in this chapter, the Catholic tradition would prohibit any experimentation on embryos for the purpose of germline modification that would cause serious harm or death. The current methods of experimentation seem to involve technical difficulties that could lead to the death or serious harm of embryos, and thus these methods would be prohibited, even though in principle germline modification might be approved of as a goal. In addition, I am not convinced that all the moral arguments in favor of these therapeutic efforts are completely satisfying as they stand.[65] Consequently, the ethical conclusion I reach is that at the present time we should not attempt to perform these therapies on either gametes or on fertilized ova before transferring them to their mothers' wombs. However, this chapter has, I hope, advanced the debate about germline therapy from a theological perspective. I would argue on theological grounds that once the scientific, public policy, and moral difficulties can be resolved, we may cautiously move forward with this type of genetic therapy. In other words, based on my theological interpretations of both the fivefold Christian themes and their anthropological counterparts and the six central themes articulated by the Roman Catholic tradition, I conclude that these therapies are not in principle as a goal fundamentally contrary to God's creative and redemptive purposes. To use them is not necessarily to arrogate to ourselves various functions and tasks that properly belong only to the divine. If developed and applied responsibly, these genetic interventions neither usurp God's rights over creation nor do they represent improper attempts to play God. On the contrary, I consider these therapeutic technologies as goals that have the potential of becoming mediations of or participations in God's redemptive activities toward humanity. Consequently, their use for the moral ends of preventing or curing serious genetic diseases in the germline can be a means of properly exercising human responsibility.[66]

## Notes

1. For an interesting empirical study and analysis of how the attitudes and beliefs of some religious leaders function in their judgments about genetic research and the Human Genome Project, see Kinh Luan Phan, David John Doukas, and Michael D. Fetters, "Religious leaders' attitudes and beliefs about genetics research and the Human Genome Project," *Journal of Clinical Ethics* 6 (1995): 237–246.

2. I would argue that the functional equivalents of theological beliefs also partially comprise the moral context in which secular judgments on genetic manipulation are formulated.

3. Thomas F. Lee, *The Human Genome Project: Cracking the Genetic Code of Life* (New York: Plenum, 1991), p. 295.

4. W. French Anderson, "Genetics and human malleability," *Hastings Center Report* 20 (1990): 21–24.

5. In the end, the President's Council on Bioethics argues that this distinction between therapy and enhancement is at least problematic, and its use "can often get in the way of the proper ethical questions." I would argue that the distinction does have some validity, although I recognize that there are conceptual difficulties with its use. For the purposes of this chapter, however, I use the distinction to analyze germline therapeutic interventions whose intent is either to cure or to prevent disease. See President's Council on Bioethics, *Beyond Therapy: Biotechnology and the Pursuit of Happiness* (Washington, DC: Government Printing Office, October 2003), pp.13–16, at 16.

6. Lee, *Human Genome Project*, p. 183.

7. Burke K. Zimmerman, "Human germ-line therapy: The case for its developments and use," *Journal of Medicine and Philosophy* 16 (1991): 595.

8. Brian V. Johnstone, C. SS. R., "La tecnología genética: perspectiva teológico-moral," *Moralia* 2 (1989): 301.

9. For a detailed description of the various techniques involved in germline manipulation, see Marc Lappé, "Ethical issues in manipulating the human germ line," *Journal of Medicine and Philosophy* 16 (1991): 621–639, at 622–623.

10. For example, see W. French Anderson, "Human gene therapy: Why draw a line?" *Journal of Medicine and Philosophy* (December 1989): 681–693 and Anderson, "Genetics and human malleability."

11. For example, see Mary Carrington Coutts, "Scope Note 24: Human gene therapy," *Kennedy Institute of Ethics Journal* 4(1994):69–72; and Maurice A. M. de Wachter, "Ethical aspects of human germ-line gene therapy," *Bioethics* 7(1993): 167–170.

12. For example, see Ronald Cole-Turner, *The New Genesis: Theology and the Genetic Revolution* (Louisville, KY: Westminster John Knox, 1993), pp. 70–79; J. Robert Nelson, "The role of religions in the analysis of the ethical issues of

human gene therapy," *Human Gene Therapy* 1 (1990): 44–47 and Nelson, *On the New Frontiers of Genetics and Religion* (Grand Rapids, MI: Eerdmans, 1994), pp. 171–183.

13. For example, see the President's Commission for the Study of Ethical Problems in Medicine and Biomedical and Behavioral Research, *Splicing Life: The Social and Ethical Issues of Genetic Engineering with Human Beings* (Washington, DC: U.S. Government Printing Office, 1982), pp. 77–78; John Paul II, "The Ethics of Genetic Manipulation," *Origins* 13 (1983): 388; Catholic Health Association of the United States, *Human Genetics: Ethical Issues in Genetic Testing, Counseling, and Therapy* (St. Louis: Catholic Health Association of the U.S., 1990), p. 20; and W. French Anderson and Theodore Friedmann, "Gene Therapy: Strategies for Gene Therapy," in Warren T. Reich, ed. *Encyclopedia of Bioethics*, rev. ed. (New York: Simon & Schuster Macmillan, 1995), vol. 2, p. 911. For a position that does not draw a line between morally permissible and impermissible forms of genetic manipulation according to the therapeutic–nontherapeutic distinction, see Nicholas Agar, "Designing babies: Morally permissible ways to modify the human genome," *Bioethics* 9(1995):1–15.

14. Alex Mauron and Jean-Marie Thévoz, "Germ-line engineering: A few European voices," *Journal of Medicine and Philosophy* 16(1991): 652.

15. For example, see the President's Commission, *Splicing Life*, pp. 46–47; and Mauron and Thèvoz, "Germ-line engineering," p. 652.

16. For example, see the Bishops' Committee for Human Values, "Statement on Recombinant DNA Research," in Hugh J. Nolan, ed., *The Pastoral Letters of the United States Catholic Bishops* (Washington, DC: National Conference of Catholic Bishops, 1983), vol. IV, p. 202; and Ian G. Barbour, *Ethics in an Age of Technology* (San Francisco: HarperCollins, 1993), p. 197.

17. Eric T. Juengst, "Germ-line therapy: Back to basics," *Journal of Medicine and Philosophy* 16 (1991):587–592.

18. Also see A. P. Cole, J. Duddington, I. Jessiman, and J. Williamson, "The human genome project and gene therapy," *Catholic Medical Quarterly* 42 (1992): 27.

19. Karol Wojtyla (John Paul II), *Love and Responsibility*, trans. H. T. Willetts (New York: Farrar, Straus & Giroux, 1981), p. 246.

20. John Paul II, "Biological research and human dignity," *Origins* 12(1982): 342.

21. John Paul II, "The Ethics of Genetic Manipulation," p. 388.

22. International Theological Commission, "Communion and Stewardship: Human Persons Created in the Image of God," §90. Available at http://www.vatican.va/roman_curia/congregations/cfaith/cti_documents/rc_con_cfaith_doc_20040723_communion-stewardship_en.html (accessed May 7, 2007).

23. Paul Ramsey, *Fabricated Man: The Ethics of Genetic Control* (New Haven, CT: Yale University Press, 1970), p. 138.

24. Bruce R. Reichenbach and V. Elving Anderson, *On Behalf of God: A Christian Ethic for Biology* (Grand Rapids, MI: Eerdmans, 1995), p. 55.

25. President's Commission, *Splicing Life*, pp. 54–56.

26. James F. Keenan, S. J., "What is morally new in genetic manipulation?", *Human Gene Therapy* 1(1990):290.

27. See Lisa Sowle Cahill, " 'Playing God': Religious symbols in public places," *Journal of Medicine and Philosophy* 20(1995):342; and Allen Verhey, " 'Playing God' and invoking a perspective," *Journal of Medicine and Philosophy* 20 (1995): 347–364.

28. Charles E. Curran has argued that these five mysteries constitute the horizon or stance of Christianity. See Charles E. Curran, *New Perspectives in Moral Theology* (Notre Dame, IN: Fides, 1974), pp. 47–86. I have argued elsewhere that the horizon of Christian religious intentionality is one of the elements that constitutes the specificity or uniqueness of Christian ethics. See James J. Walter, "Christian ethics: Distinctive and specific," *American Ecclesiastical Review* 169 (1975):483–484.

29. A few years ago Kenneth Vaux helpfully used a similar framework based on these Christian themes to discuss the theological foundations of medical ethics in general. See Kenneth Vaux, "Theological Foundations of Medical Ethics," in Martin E. Marty and Kenneth L. Vaux, eds. *Health/Medicine and the Faith Traditions* (Philadelphia: Fortress, 1982), pp. 215–228.

30. There are, of course, a number of other background issues that function as presuppositions to moral judgments on this type of genetic therapy. The following list is merely a sample of such issues: the goals and limits of medicine [Eric T. Juengst, "The NIH 'points to consider' and the limits of human gene therapy," *Human Gene Therapy* 1(1990):426, 431], the meaning of suffering and illness [Lucien Richard, O. M. I., *What Are They Saying About the Theology of Suffering?* (New York: Paulist, 1992)], attitudes about genetic disabilities [Barbour, *Ethics in an Age of Technology*, p. 196], and the relation between science and theology [Zachary Hayes, O. F. M., *What Are They Saying About Creation?* (New York: Paulist, 1980), 7–20; James M. Gustafson, *Ethics from a Theocentric Perspective* (Chicago: University of Chicago Press, 1981), vol. 1, pp. 251–259; and J. Robert Nelson, *Human Life: A Biblical Perspective for Bioethics* (Philadelphia: Fortress, 1984), pp. 167–170].

31. Bernard Häring, *Medical Ethics* (Notre Dame, IN: Fides Publishers, 1973), p. 47.

32. See Josef Fuchs, S. J., *Christian Morality: The Word Becomes Flesh* (Washington, DC: Georgetown University Press, 1987), pp. 39–61 and Jan Jans, "God or man? Normative theology in the Instruction *Donum Vitae*," *Louvain Studies* 17 (1992): 56–63.

33. Philip Elmer-Dewitt, "The genetic revolution," *Time*, January 17, 1994, p. 48.

34. Ian G. Barbour, *Issues in Science and Religion* (New York: Harper & Row, 1966), p. 449.

35. Cole-Turner (*The New Genesis,* pp. 86–93) adopts a view similar to this model, except that he judges the disorder of disease in nature to be a moral disorder (p. 89). In my position, the disorder that exists in nature is not moral but ontic or premoral, i.e., it is prior to the moral order. Nonetheless this disorder is truly evil because it causes great human suffering. Consequently, the disorder needs to be overcome or corrected as far as possible, i.e., other things being equal, through human efforts.

36. Other interpretations of the *imago Dei* have also been proposed. For example, John Paul II ("The Ethics of Genetic Manipulation," p. 389) has recently described our role in creation as that of a king.

37. Thomas A. Shannon, *What Are They Saying About Genetic Engineering?* (New York: Paulist, 1985), p. 21.

38. In specifying our responsibilities toward genetic manipulation, D. Gareth Jones ["Manipulating human life: The ambiguous interface between science and theology," *Colloquium* 26 (1994): 26–28 and "Making human life captive to biomedical technology: Christianity and the demise of human values," *Update* 11 (1995): 3–4] adopts the stewardship model of the *imago Dei*, but then he goes on to define our responsibilities in terms of the co-creatorship model.

39. James Gustafson (*Ethics from a Theocentric Perspective*, vol. II, p. 294) has preferred to describe our role in creation as participants rather than co-creators. He argues that the divine continues to order creation, and we can gain some insight into God's purposes by discovering these ordering processes in nature.

40. Philip Hefner ["The Evolution of the Created Co-Creator," in Ted Peters, ed., *Cosmos as Creation: Theology and Science in Consonance* (Nashville, TN: Abingdon, 1989)] is the originator of the phrase "created co-creator." He used the adjective "created" to indicate that we are created beings and thus not creators in the same way that God is, i.e., we do not create *ex nihilo*.

41. Ted Peters, " 'Playing God' and germline intervention," *Journal of Medicine and Philosophy* 20 (1995):377–379; and Ann Lammers and Ted Peters, "Genethics: Implications of the Human Genome Project," in Paul T. Jersild and Dale A. Johnson, eds., *Moral Issues and Christian Response* (New York: Harcourt, Brace Jovanovich, 1993), p. 302.

42. Daniel Callahan, "Living with the new biology," *Center Magazine* 5 (1972): 4–12.

43. Shannon, *What Are They Saying About Genetic Engineering?*, p. 37.

44. W. Norris Clarke, S. J., "Technology and Man: A Christian Vision," in Ian G. Barbour, ed. *Science and Religion: New Perspectives on the Dialogue* (New York: Harper & Row, 1968), pp. 287–288.

45. For a helpful article that assesses the applicability of the slippery-slope argument to human gene therapy, see Tony McGleenan, "Human gene therapy and slippery slope arguments," *Journal of Medical Ethics* 21 (1995): 350–355.

46. For example, see Joseph F. Fletcher, "Ethical aspects of genetic controls: Designed genetic changes in man," *New England Journal of Medicine* 285 (1971): 776–783.

47. It is possible in this view to hold that human sinfulness may exacerbate the experience of the effects of these genetic diseases, but it is not the cause of the diseases.

48. W. French Anderson, "Genetic engineering and our humanness," p. 758.

49. Anderson, "Genetic engineering and our humanness," p. 759.

50. Professor the Reverend Canon G. R. Dunstan, "Gene therapy, human nature and the churches," *International Journal of Bioethics* 2 (1991): 236.

51. See John Paul II, "Biological Research and Human Dignity," p. 342 and "The Ethics of Genetic Manipulation," p. 388; Johnstone, "La tecnología genética: perspectiva teológico-moral," pp. 307–308.

52. Paul Ramsey, *Fabricated Man*, p. 133.

53. For a further analysis of the Human Genome Project and its relation to the concept of human nature, see Max Charlesworth, "Human Genome Analysis and the Concept of Human Nature," in Derek Chadwick et al., eds., *Human Genetic Information: Science, Law and Ethics* (New York: John Wiley & Sons, 1990), pp. 180–188.

54. See Barbour, *Ethics in an Age of Technology*, pp. 10–15; and Mauron and Thèvon, "Germ-line engineering," pp. 649–650.

55. Jacques Ellul, *The Technological Society* (New York: Knopf, 1964).

56. See Barbour, *Ethics in an Age of Technology*, pp. 4–8; and Mauron and Thèvon, "Germ-line engineering," p. 650.

57. Pierre Teilhard de Chardin, *The Phenomenon of Man* (New York: Harper & Brothers, 1959).

58. See Scott B. Rae and Paul M. Cox, *Bioethics: A Christian Approach in a Pluralistic Age* (Grand Rapids, MI: Wm. B. Eerdmans, 1999), pp. 118–127.

59. This position is more informed by St. John's Gospel than St. Paul's epistle. In John 9:1–3, the evangelist writes, "As he went along, he saw a man who had been blind from birth. His disciples asked him, 'Rabbi, who sinned, this man or his parents, for him to have been born blind?' 'Neither he nor his parents sinned,' Jesus answered, 'he was born blind so that the works of God might be displayed in him'."

60. Ronald Cole-Turner, *The New Genesis*, pp. 80–97.

61. Harvey Cox, "Evolutionary Progress and Christian Promise," in Johannes B. Metz, ed., *Concilium* vol. 26, *The Evolving World and Theology* (New York: Paulist, 1967), pp. 35–47.

62. Cox, "Evolutionary Progress and Christian Promise," in Metz, ed., *Concilium*, p. 38.

63. Cox, "Evolutionary Progress and Christian Promise," in Metz, ed., *Concilium*.

64. Ramsey, *Fabricated Man*, p. 27.

65. For example, I am not convinced that the fourth argument for this type of intervention listed by Juengst ("Germ-line therapy: Back to basics," p. 590), which is concerned with the respect that is due for parental autonomy, is entirely adequate as it stands. To be sure, parental autonomy is an important value, but it is not clear to me that this value should necessarily override other moral considerations.

66. An earlier version of this chapter was published by *New Theology Review* and is used here with permission.

# 7

# Germline Genetics, Human Nature, and Social Ethics

Lisa Sowle Cahill

The distinctive moral aspect of genetic modification of the germline is the physical heritability of the changes they produce. This chapter considers such changes in light of their probable social significance and impact, with special attention to the consumer-oriented marketing of genetic enhancements. The relation of genetic intervention to a concept of human nature is also addressed, granting that such a concept is difficult to define. I will defend a concept of human nature that includes rationality, free will, and sociality and invoke that concept to advocate restraints on genetic manipulation of the germline.

Both germline and somatic interventions have an impact on human nature insofar as they change or adjust human traits—physical, behavioral, and social. However, inheritable genetic changes have a different impact on human nature than those that directly and physically affect only one individual or a series of individuals. Germline modifications have a greater impact on human social relationships and social institutions continuing over time than somatic modifications. Both positive and negative effects of interventions will be passed on to succeeding generations. To the extent that the genetic benefits and risks of the original modification are accessible or allocated according to social or economic status, the allocating patterns of social relationships may be perpetuated among descendents whose genetic traits reflect the social positions of their progenitors. In other words, the privileged can access therapies and enhancements that reflect their prestige and power, and these will be passed on to their genetic heirs. Assuming that these genetic traits confer advantages, the advantages and the social success enabled by them will also be passed on.

This situation is analogous to the inheritance of social advantage and disadvantage by other means, such as education, wealth, health care, cultural competence, and social connection. However, genetic inheritance and social inheritance are not identical in character or effects. Genetic inheritance is less susceptible to subsequent resistance, reinforcement, renegotiation, or dissipation by descendants, or by subsequent social dynamics, than is socioeconomic inheritance. This can be substantiated by reference to studies of behavioral genetics.[1] Traits usually associated with personality and character should theoretically be the most amenable to social influence and the least affected by biological determinants. However, research, including studies of twins, indicates that while environment plays an important role in the expression of traits, physical and behavioral tendencies associated with genetic inheritance are surprisingly resurgent. Thus, while inheritable genetic modification does not guarantee a particular result, either immediately or intergenerationally, it can certainly tip the balance in its favor.

I do not believe that it is possible to conclude that germline intervention is intrinsically wrong. However, it is contingently wrong. Germline therapy is attended by uncertainties and risks that are significant enough to create a moral presumption against its use in any case at present. However, germline therapy for disease could be morally permissible or even obligatory if safety issues were resolved. Germline enhancement is much more problematic given not only its risks but also the difficulty of defining what constitutes enhancement, and given the social control of its advantages by elites. I conclude that it is incumbent on theological and other bioethicists to denounce and resist the marketing of genetic enhancements that is already well on its way to realization as a common social practice in privileged cultures.

## What Is "Human Nature"?

From a theological perspective, human nature is a normative concept that is frequently associated with the biblical symbol, "image of God." Historically, it has been developed by theologians and philosophers primarily with reference to what are considered to be the uniquely human characteristics of reason and free will. Reason and freedom are also key to western philosophical anthropologies, e.g., to Plato's and Aristotle's definitions of the soul, of virtue, and of the good society.

As rational and free, human beings are capable of speech and intentional, conscious action, as well as the empathetic and reflective understanding of the thoughts and emotions of others that is the basis for compassion and altruistic social action. Reference to the human body as constituted by certain characteristic needs, capabilities, functions, and purposes has also been part of theological and philosophical consideration of the meaning of human nature. Human embodiment helps define the conditions and contexts of the exercise of reason and freedom and provides parameters or guidance for what that exercise should be. For example, the physical conditions of human life, growth, reproduction, and production should be respected and protected for oneself and one's own group and for others, and by means of consistent and extensive social practices.

The interdependence in human nature of physicality with freedom and reason is one of the main sources of ambiguity in the concept of human nature. Both philosophy and common sense tend to regard human nature as static and unchanging, and to assume a clear demarcation between the natural and the artificial. However, even the "natural" human body as we know it today, with its shared DNA and its relative constancy of form and function across cultures, is the result of evolutionary forces resulting partly from human behavior, such as migration, warfare, and intermarriage. Human beings are an active part of a constantly evolving system.[2] "The unique value of humanity—its dignity—lies in its power of self-transcendence, of being other than the natural given."[3]

Human nature is intrinsically open—but does this mean that it has no recognizable essence or identity over time, or that its path into the future cannot be guided by norms? No. It is true that humanity is a historical reality with a diachronic identity that is neither completely unique in comparison with the capacities of other mammals, nor definitively settled in all its particulars. Nevertheless, the remains and artifacts of the earliest humans—in significant continuity, it must be said, with protohumans and other primates—share recognizable similarities of physical form and function, reproductive activities, productive activities, and even cultural activities such as art and religion, with humanity as we know it from the beginnings of recorded history onward.

Ludwig Siep argues that "the human body as a whole . . . is the basis and point of reference for our social rules." The body is a common basis on which all cultures appreciate certain values that ground "criteria for what we owe to other people."[4] These values include food, shelter, safety from physical violence, conditions favoring family formation, education in the means of productive labor, political systems that coordinate the needs of many, and even art and religion, insofar as they interpret the place of the human in its natural and social environment and give meaning to suffering, death, ecstasy, and love. In other words, insofar as all human experience is embodied, the body provides the indispensable baseline for understanding and moral agency within the human condition. According to Siep, "we should regard the 'traditional shape' of the human body as a common heritage, not simply as property and a tool of its owner who can do with it whatever she or he wants." The heritage can be changed for future generations only to avoid "heavy suffering," as defined on a common " 'evaluative view' of the human body."[5]

If the human body, with its needs and claims, has a recognizable consistency over time that provides guidance for human perceptions of value, so too does human society. Social organization, politics, culture, and religion are all interdependent from, and even arise from, the human experience of embodiment. Psychosocial experiences such as rationality, freedom, affectivity, and intersubjectivity are all mediated by the body, the senses, and the means of communication furnished by the body and by signs that are sensually perceptible. Virtually every society provides for and organizes production, reproduction, the political organization of members, religious experiences, and cooperation or conflict with other societies. The social organization of human relationships is just as essential to human nature as are reason, free will, and embodiment.

Values and normative criteria follow from the conditions of social life that support human well-being: security in relation to basic physical needs, such as food and water; stable and productive internal organization; and protection from external enemies. Although the specific forms guaranteeing these conditions will be culturally diverse, all cultures recognize their essential desirability and seek to secure them. Violent conflict over the means to do so is universally seen as an evil; nevertheless, often those with the means to do so use violence to seek dominance over social goods.

The critique has been made of the purveyors of modern genetic science that this profoundly relational and social view of the human has been replaced by one that overemphasizes physicality. In this latter view, medical and technological modifications of the body are seen as potential solutions to human and social problems, and even to suffering and mortality as such. The wider social contexts and significance of the body, and of the human as embodied in social relationships, are cut off from the assessment of biotechnology. According to Robert Song, this reductionist approach gives "the explanatory priority" to biology. It assigns genetic makeup much too important and essential a role in defining social identity and thus marginalizes social and environmental factors that contribute to the malleability of traits and that can be affected by social reform.[6] Control of human nature through specific technological interventions in the body is seen as a way of controlling fate and necessity, of eliminating the inevitability of suffering, and even of avoiding death indefinitely. This accounts for the tremendous cultural popularity of genetic science and its promises, at least among those who expect to be able to afford them.[7]

Philosophically and politically, genetic interventions are often defended in terms of the rational control of risks and benefits and the right to autonomy and free choice. As I have maintained, however, reason and freedom are socially conditioned, expressed, and structured. Their purpose and effects are social. Their acceptable use or expression can therefore be subjected to criteria of social well-being and equity. An essential norm of human activity as reasonable, free, embodied, and social is justice, understood as the inclusive and equitable participation of all human beings in the communities to which they belong and in whose goods and benefits they are entitled to share.

The definition of justice upon which I rely, and which I will apply in the area of germline modification, is procedural, substantive, social, and global. Justice requires procedures by which all social members can participate in establishing practices and institutions that affect their welfare and that of their communities. Justice requires access to the basic human goods necessary for human life, well-being, and society. Justice refers to and includes patterns of social relationships and institutions that allow individuals and groups to be related to one another consistently at distances of time and space. Justice as participation, as sharing in basic goods, and as social or political is a global norm or ideal, applying to all peoples or cultures. While the specific forms of participation, the prioritization and distribution of basic goods, and the structuring of society and politics can and do vary immensely among eras and cultures, this does not obviate the ability and responsibility to hold up justice as a norm for the relationships among peoples and societies around the globe. It does require that specifications of what justice requires be approached inductively, contextually, dialogically, and provisionally, and in proportion to their particularity.

**Some Theological Perspectives on Human Nature**

Human nature is understood normatively by philosophers and theologians in that the concept is held up as an ideal of human identity and behavior, rather than as a definitive or comprehensive description of the existential realities of members of the species. The ideal is dynamic, although—like the body—it is neither discontinuous with past forms nor random in its forward movement. Biblically and theologically, as well as historically and experientially, this dynamic human nature is social. Human relationships are as constitutive of human identity as the characteristics of reason and freedom.

Theologians often define the normatively human by appealing to the biblical story in which humans were created in the image of God (Genesis 1:27). In twentieth-century theology and ethics, this concept functions as "a root metaphor for the Christian understanding of the human person, the religious way of grounding the inviolability of human dignity, and the basis for defending the human rights of all persons."[8] Similarly in the modern liberal political tradition of human rights, theological invocations of the image of God tend to focus on the individual person as the focus of inviolability and hence to define the image in terms of individual characteristics. The morality of genetics and genetic engineering is often pursued in the same vein by studying whether they entail violations of human characteristics or rights that inhere in or can be defined with reference to individual humans. Even when the agent of genetic manipulation is presented in the plural, as "we," it is often the case that "our" actions are construed in a mode similar to that of individual decision making abut individual bodily capacities or states. For example, Paul Ramsey denounces "genetic control" in the form of in vitro fertilization in the following terms: "God created nothing apart from his love; and without the divine love was not anything made that was made. Neither should there be among men and women (whose man-womanhood—and not their minds or wills only—is in the image of God) any love set out of the context of responsibility for procreation, any begetting apart from the sphere of love."[9]

In a recent book on genetics, Thomas Shannon and James Walter note that the *imago Dei* is typically construed in one of two different ways as a model for responsible human action: stewardship or co-creatorship. The stewardship model "accentuates the fact that humans are entrusted with responsibility for conserving and preserving creation," and tends to place limits on human freedom to alter what the divine has created."[10] The co-creator model holds that "we do model the divine in our capacity to create," and that "because we cocreate with the divine, we have greater freedom than in the previous model to intervene into our genetic material."[11] The co-creator metaphor reflects the fact that since both freedom and ongoing relationships are part of human nature, it is impossible to tie down a specific definition of human nature that is not also subject to the malleability entailed by the definition of the human itself. However, this metaphor is problematic when it does not firmly refer human co-creation to the distinctively human condition, including not only reason and freedom but also finitude, fault, and the social effects of human actions in, through, and on the body.

The conclusion does not follow from freedom and openness that it is impossible to stipulate moral and social norms based on human nature considered holistically. The nature of the human must be defined so as to include social relationships and institutions. Biotechnology and genetics, for example, must be evaluated in light of their social effects. Both the stewardship and the co-creator model see human beings in relation to God, other humans, and the natural world. Yet the moral focus in both is on the character of the human agent's intelligence and freedom, responsibly used, rather than on the relationships within which they arise, or on the institutions that give relations and agency their collective forms.

In some recent interpretations of Genesis, it is relationality itself that is the image of the divine in the human, and relationship is the way in which the image is fulfilled. In these interpretations, relationships have a corporate dimension, and relationship is the condition of possibility of the human characteristics of reasonableness and free choice. Douglas John Hall contrasts a "substantialist" and a "relational" strand of historical interpretation of the image of God. The former line of thought tends to emphasize rationality and to devalue the human body and the material, physical conditions of human existence. Hall prefers a line that links the image of God with relationality and sees humanity as created to be a "being-in-relationship." Essential human nature cannot be known by looking at individuals in isolation from one another, but only "by considering human beings in the context of their many-dimensioned relationships."[12] Intelligent consciousness and the freedom to shape and direct one's desires exist for the purpose of entering into relationship. "Relationship is the essence of the creature's nature and vocation."[13]

Biblical scholar Claus Westermann confirms that humans' relationships to one another and to God are part of the divine image in humanity. Humanity in the creation narrative has a collective meaning; the narrative is concerned with the human race or the species.[14] Thomas Mann identifies two literary devices uniting the book of Genesis that also refer to humanity's relational nature as the primary context for human responsibility before God. These are the "generations" formula, which occurs eleven times in Genesis, and the divine "promises" of blessing, which begin immediately after the creation of male and female in God's image (1:28) and are eventually extended to "all the families of the earth" through Abraham (12:3).[15] These promises include children, land, nationhood, and a communal covenantal relation to one another and to God. Image of God language can also be used to affirm that individual humans are members of the community of relationship and thus to secure for all a share in the community's goods and benefits, including genetic therapies.[16]

A relational view of human nature as dynamic and forward moving is illustrated in the biblical promises of fulfillment through relationships and a future that moves across generations. Intelligence and freedom are capacities of the human through which individuals and communities join in social interaction over time. Dynamic identity and relationality are not always realized positively, however. Relationships and collective identities can be distorted by perspectives and agendas that violate the norms of justice or the common good as defined earlier. The biblical and theological category for such distortions is sin. James Keenan observes that the goal of perfection seen from the standpoint of a subject intrinsically and constitutively in relationship to others and to God has a referent that goes beyond biology. Humans are within nature and are capable of transforming it, but human finitude and an orientation to that which transcends human goods and goals provide horizons and parameters within which perfection should be sought.[17] The agenda of co-creation is not immune to distortions and should not be promoted in the absence of careful consideration of the criteria to which it should be subject.

Julie Clague enumerates the requirements of the common good in the context of genetics in terms of several moral concerns. The social good or public interest must be protected as well as the rights of individuals. The common good also implies an equitable distribution of the benefits available in societies.[18] Theologically, the concept of the common good is qualified by the virtue of solidarity and a commitment to prioritize the welfare of those who are most marginal in any society or situation. Biblically rooted in the prophets of the Hebrew Bible and the teaching and example of Jesus, the so-called preferential option for the poor is the keynote of liberation theology. The relationships that make up the common good ever more urgently demand a widening of vision to include global society, and the application of criteria of the public interest, individual protections, and equitable access to benefits at the global level. At the global level, the preferential option for the poor exerts a strong moral pressure against marketing genetic advantages for the welfare of the wealthy.

**Genetic Interventions: Therapy and Enhancement**

Put simply, the aim of therapy is to alleviate disease, dysfunction, or pathology, while enhancement is meant to improve on normal human traits. These definitions are fraught with ambiguity, however. An initial problem is defining the parameters of normal constitution or function and deciding what line must be crossed to constitute pathology, e.g., in the case of height, weight, or longevity. A second problem is that interventions originally developed to compensate for or alleviate deficiencies can also be used to build on and increase a normal capacity, e.g., muscle strength or memory. Another problem is that criteria of normality or health may be raised with the frequency and success of interventions, as might be the case if a tendency to treatable conditions such as diabetes, poor vision, or learning disabilities could be eliminated through germline therapy. Finally, while all these definitions have a basic reference to the health and functioning of the human body, and while at a fundamental level those terms have a stable meaning, cultural interpretations of what is desirable or undesirable in a certain area become more important at the blurry edges of each category.

A normative ethics of germline modification cannot be derailed by asserting, as has one philosopher, that "there is no set of fixed properties (either of the ideal human being or the average or normal) which forms what humans essentially are," or that "the open-ended character of human nature only supports a reasoned choice of self-transformation." The same thinker asserts that "there may be good reasons for restricting genetic enhancement (like social distributive injustice), but the argument from human nature cannot serve as one of them."[19] It is not necessary to stipulate a complete set of fixed properties in order to defend the proposition that there is a recognizable and defining similarity among human beings past and present, physically, mentally, and culturally. Moreover, the fixed properties approach to nature reflects the modern fixation with individuality that truncates the concept of nature by leaving out sociality. This mistake is exemplified in the invalid distinction between human nature and the requirements of human social well-being (social distributive justice). Without some sense of the personal and social goods appropriate to human flourishing, how would a reasoned approach to the transformation of the individual be possible at all? The debate about the ethics of genetic therapy assumes and requires a reference to a normal state of human health. The debate about enhancement requires a reference to humanity's social nature and how it may be affected by genetic manipulation of the germline.

Without some idea of normal human functioning, the practice of medicine would be completely rudderless. Although an expansive definition of health, like that of the World Health Organization,[20] can be an important strategy in calling international attention to all the social conditions of health, it cannot serve as the foundation for more specific bioethical analysis. Indeed, while academic philosophers may debate whether human health is a meaningful concept, such ponderings seem irrelevant if not frivolous in the face of the UN Millennium Declaration, which targeted worldwide rates of maternal and infant death, the global ravages of AIDS, and the "diseases of the poor," or the Asian tsunami disaster of December 2004 that galvanized worldwide relief efforts.

The authors of a major collaborative study of new genetic technologies, *From Chance to Choice*, rightly presume the self-evidence of "some very basic characteristics" of human beings, some "primary goods," that allow the flexible and successful "pursuit of a broad range of human projects in a diversity of social environments."[21] One of the authors, Norman Daniels, defends a narrower rather than an expansive definition of health and disease, defined by conformity with or "*deviations from the normal functional organization of a typical member of a species*." Normal functional organization permits members of the human species "to pursue biological goals as social animals; our various cognitive and emotional functions must be included," as must mental health, despite the difficulty of defining it precisely.[22]

As the collective authors assert, "To the extent that the genetic factors that contribute to these can be accurately identified and subjected to safe and effective human control, there is a strong *prima facie* case for undertaking efforts to reduce the impact of inequalities in their distribution."[23] Although no society can guarantee equality of health or genetic assets, it may at least be possible, desirable, and just to aim at something like "a 'genetic decent minimum'."[24] At the very least there is a presumption that justice requires the prevention or amelioration of diseases that seriously limit opportunities.[25]

*From Chance to Choice* also contemplates the possibility that it might be appropriate to use genetic means to combat serious natural inequalities that do not actually amount to diseases, such as low intelligence, as long as these means are not available solely on the basis of ability to pay.[26] If genetics could provide socially advantageous reinforcement of capacities like memory, concentration, and resistance to common physical and mental illnesses, these could be justifiable, even though they might "ratchet up" the standards of normality. If instituted, justice would require that such improvements be part of a basic genetic package provided by universally available health care and not be subject to market allocation, exacerbating existing inequalities.[27]

Such suggestions take us across the permeable boundary between therapy and enhancement. I maintain that the major ethical objection to germline enhancement is accessibility, a point to which I will return. However, even before we get to the distribution question, it can be argued that genetic means to individual improvement are intrinsically problematic from the standpoint of their effect on the cultivation of character and the human ability to contend with adversity. "By offering us an easy way to achieve the end, the new means cheat us of the value to be found in the old means. There is, after all, a glory and a dignity in human accomplishment attained the 'old-fashioned way,' through sweat and struggle, sometimes against great odds."[28]

Moreover, two different means to an end may accomplish the same result, but only in respect to one aspect of the outcome. Other morally significant consequences can be different. Ronald Cole-Turner uses the example of a state of ecstatic self-transcendence achieved by pharmacology or prayer.[29] The magnitude of the changes accomplished by different means can also be different. For example, germline genetic engineering is much more precise in achieving desired traits than more chancy methods like "matchmaking," and certainly more so than education.[30] A shortcut to an end can even undermine the social practice that makes the trait valuable in the first place, as is perhaps most obvious in the case of athletic competition.[31] It is obvious to most commentators that the desirability of the traits selected is highly interdependent with cultural views, particularly where genetically linked behavioral traits are concerned. It would not be surprising if those with the resources to obtain genetic enhancement services for themselves and their offspring selected for traits that enhanced competitiveness in modern western societies, rather than for traits like emotional sensitivity or compassion. Selecting for culturally valued traits can also amount to complicity with cultural definitions of normality that are biased and discriminatory.[32]

Even more problematic is the probability that working on a biological solution to a human and social problem or need will discourage attention to the complex underlying social conditions of human suffering. The more successful genetic technology is in alleviating stressful responses to adverse environments, the less inclined society may be to change the hostile and unfair conditions to which some humans subject others.[33] Resort to genetic solutions also affects the human sense of self, of an authentic human life, and of personal responsibility. "Opting only for the pharmacological, mechanistic response lends itself to our thinking of ourselves [and our children] more and more in mechanistic terms—and less and less in terms of being responsible agents [and educators and parents]."[34] Thomas Szasz believes that modern Americans have created a "pharmacracy," in which "the idiom, imagery, and technology of medicine" have been extended into almost every area of human concern, and the rule of medicine has become the new rule of law.[35]

All these considerations are strong indicators of the inadvisability on moral grounds of genetically manipulatory the germline. No one argument conclusively proves that such manipulation is intrinsically wrong. Nevertheless the case against it is compelling, particularly in view of the social nature of human beings and the normative force of a concept of justice focused on the common good.

## Germline Interventions, Justice, and the Market

The apparently easy accessibility of biotechnological solutions to social problems has posed some important initial questions about how well such solutions comport with a notion of the common good as inclusive and participatory. An even bigger challenge to the common good is raised by the virtual certainty that access to biotech solutions will favor the wealthy and influential. Even the elimination of deleterious genes could lead to further inequities, because the health of successive generations of the wealthy contributes to their ongoing competitive edge. This effect might be diluted by the facts that remedies for disease will be available to many people under health insurance, and natural selection will tend to foster the dominance of healthy over deleterious genes in the general population. In comparison with germline therapies, germline improvements are much more worrisome. Enhancements will be chosen with greater frequency by healthy consumers and hence will multiply inequities much more significantly than therapies for disease.

Given the current state of health care access in this country and globally, the idea that gene modification of any sort will be available to all as part of a basic package of health care is ludicrous. It is well known that there are 45 million uninsured people in the United States and that the numbers are growing. Meanwhile, the growth of health expenditures in the United States actually slowed in 2003 (to 1.7 trillion) for the first time in seven years, although they became a bigger percentage (15.3) of the gross domestic product than ever before. Why this change in the trend of total spending? According to a 2005 annual government report, the slower growth pace in 2003 was driven by a slowdown in the availability of public dollars, specifically, financial constraints on the Medicaid program and the expiration of supplemental funding provisions for Medicare services. Out-of-pocket health care payments by individuals increased at a faster rate than the overall national health spending owing to rising numbers of uninsured people and cutbacks on coverage by employers.[36] Drug sales also increased faster than the rate of overall national spending on health. What might this portend for the future of genetic innovations in health care?

Against this backdrop it is quite evident that universally provided equitable health care extending to genetic technologies will be entirely out of reach for the foreseeable future. For that matter, basic health care is a much more urgent need than genetic innovations. That is true for the uninsured and inadequately insured in this country, and is a truism for the developing world. According to a U.S. physician with experience in bioethics, theology, and international health issues, "malnutrition and infectious disease, including typhoid, malaria, dysentery, cholera and now AIDS, Ebola and SARS" are more important." "A few simple, inexpensive approaches such as immunization, simple antibiotics, and intravenous fluids with the means to administer them, are often what is most appropriate. Yet, only a few people worldwide have access to truly beneficial medical interventions."[37] For reasons like this, the Global Fund to Fight AIDS, Tuberculosis and Malaria was founded in January 2002.[38] These diseases are still far short of a world funding priority, however. In January 2005, a report from the UN Millennium Project called on recalcitrant wealthy nations, of which the United States donates the smallest percentage (0.15%) of its national income, to reinvigorate their commitment to reduce global poverty by meeting goals set in 2000.[39]

It is obvious that new germline techniques, especially enhancements, if safe, will be distributed on a market basis to those whose resources match their desires. Poor people will be forced to resort to abortion to attain children with the level of normality defined by the preferences of those with access to germline therapies. The eroding line between therapy and enhancement, and the ability of the rich to access the latter, will result in a "genetic underclass" made up of those who are dependent on federal entitlement programs for care.[40]

There is no reason not to assume that access to genetic technologies will not be channeled to consumers in much the way pharmaceuticals are today, reinforcing social disparities by disparities in access. In her scathing exposé of the drug industry, former *New England Journal of Medicine* editor Marcia Angell shows the alarming degree to which the profit motive has corrupted medical policy, practice, and research. Facing an actual downturn in innovation and in the development of new products, pharmaceutical companies are desperate to maintain their incredible profits. They achieve this by maintaining monopolies on drugs, by introducing new drugs that are little more than copies of old ones, by promoting new drugs that may be less effective than old ones, and by spending a huge proportion of their budgets hiring researchers they can control, bribing doctors, and marketing directly to consumers.[41] In the race to the bank, these companies abandon unprofitable products with little regard for the consequences for individual or public health. Serious recent drug shortages have been experienced as a result, including pharmaceuticals for premature infants, hemophilia, and cardiac resuscitation; adult vaccines for flu and pneumonia; and childhood vaccines for diptheria, tetanus, whooping cough, measles, mumps, and chickenpox.[42]

The drug companies gained unprecedented political influence in the 1980s and 90s, when their profits were at their highest. Even during the economic downturn in 2002, the profits for the ten drug companies in the Fortune 500 were more than the profits for all the other 490 businesses put together.[43] Angell reports that "big pharma" employs large-scale lobbying (often hiring former members of Congress as its representatives), creates what are ostensibly grassroots or citizens organizations to promote its cause in the media, and contributes to political campaigns on a grand scale. For example, in the 1999–2000 election cycle, pharmaceutical corporations gave $20 million in direct contributions and $65 million in "soft" money to political campaigns.[44]

Angell identifies the prescription drug benefit that was added to Medicare in 2004 as one result. First of all, the amount of money allocated by Congress for the benefit will quickly be exceeded by rising drug costs and an overly complicated scheme of administration that is designed to prevent Medicare itself from exercising any real control over it. Worse, the Medicare bill explicitly prohibits Medicare from using its influence as a potentially huge buyer of drugs to bargain for reduced rates. As a result of the bill, the market for drugs will expand, but so will the prices. As prices rise, so will deductibles and copayments for seniors, while other Medicare benefits may very well decrease.[45] Taxpayers will have to absorb the cost of implementing this bill with its vastly insufficient budget allocation. In other words, they will further subsidize the drug industry.

The pharmaceutical industry also increasingly controls academic research, supposedly an objective source of data about drug safety and efficacy. University researchers and their institutions are paid to conduct trials, are promised a cut of the profits via patent and royalty agreements, and are denied information about the overall outcome of trials that are being conducted simultaneously at multiple sites. For example, about two-thirds of academic medical centers hold equity in startup companies for whom they do research, and drug companies are major benefactors of medical schools.[46] Individual researchers also form profitable relationships with such companies, serving as consultants, board members, and promoters of the company's products. Much the same is true of National Institutes of Health scientists.[47]

Lamenting the same state of affairs, Sheldon Krimsky claims that the "unholy alliance" between university scientists and the drug industry has led to the demise of science in the public interest.[48] Academic researchers can no longer be counted upon to pursue solutions to major societal problems in health and human welfare, or to honestly examine and confront the effects of new technologies. Priorities are dictated by commercial rather than social needs.[49]

The current behavior of drug companies, researchers, providers, and consumers is a good indicator of how access to and use of new genetic treatments, including germline technologies, are like to be institutionalized in the future. Mark Frankel is absolutely and stunningly correct in asserting that contrary to earlier eugenics movements or the scenario of Aldous Huxley's *Brave New World*, "the discoveries of genetics will not be imposed on us. Rather, they will be sold to us by the market as something we cannot live without."[50] Frankel points out that germline interventions for disease are virtually superfluous given the availability of preimplantation genetic diagnosis. The main use of germline manipulation will thus be for enhancement. Given cultural endorsement of reproductive freedom and the right and duty of parents to provide for their children the best future they can, enhancement is likely to be accepted widely and uncritically as an extension of means already available. While relatively few people suffer dysfunctions rooted in genetic defects, all parents want to secure advantages for their children. People in wealthy societies and social classes will do so with little hesitation and less regard for the eventual effect of their choices on the less fortunate classes of the future.

A wide range of authors, including most of those treated here, have called for public debate on these issues. Frankel concludes with a call for "the test of public discourse," based on rigorous assessment of the impact of inheritable genetic modifications, and "explicit public approval" eliminating all "backdoors, whether due to gaps in public policy or an aggressive marketplace through which IGM inches its way into our lives."[51] The need for public debate and action is certainly strong. Yet the ethical calls of academics and other theoreticians can appear weak and utopian, given the political power of the pharmaceutical industry and the complicity of government and research science with market interests. For example, the not-incorrect assertions of *From Chance to Choice* that "state action to regulate markets as they distribute the fruits of the genetic revolution is necessary," and that "the state can encourage the medical profession to reflect on its appropriate role,"[52] seem innocuous and ineffectual in the face of the evidence that Angell marshalls about the undue influence of drug profitability in determining the actions of campaigners, legislators, and political parties. Angell's book, written in an accessible and even rabble-rousing style, is aimed specifically at invigorating public protest and action. She concludes that commercial control over pharmaceuticals could be dislodged "with simple Congressional legislation," and urges, "This is where you come in. Your representatives in Congress will not deviate much from the industry script unless you force them to."[53] She also follows up with several specific and confrontational measures that ordinary citizens can use with their health care providers and congresspersons.

Theological bioethicists, likewise, need to awaken and energize, not only their colleagues and peers, but also the members of their universities, denominations, congregations, and communities. Internet venues are multiplying, with some religious organizations that are morally invested in health care taking the lead.[54] Religious leaders and theologians should also take advantage of the opportunities for grassroots and midlevel education and organization offered by their institutional structures. There is great potential in the vital and pervasive presence of churches and religiously sponsored institutions of higher education and social service agencies in the United States and abroad. Ultimately the ethical social response to genetic enhancement of the germline will require much more than theoretical analysis, whether theological or philosophical. Like other urgent demands of the common good in health and biotechnology, a genuinely ethical and adequate response to inheritable genetic interventions will require widespread and practical challenges to the economic status quo in health care research and delivery.[55]

## Notes

1. See Erik Parens, "Genetic differences and human Identities: On why talking about behavioral genetics is important and difficult," *Hastings Center Report Special Supplement* 42:1 (2004): S1–S36.

2. Kurt Bayertz, "Human nature: How normative might it be?", *Journal of Medicine and Philosophy* 28:2 (2003): 134–136.

3. David Heyd, "Human nature: An oxymoron?", *Journal of Medicine and Philosophy* 28:2 (2003): 168.

4. Ludwig Siep, "Normative aspects of the human body," *Journal of Medicine and Philosophy* 28:2 (2003): 174.

5. Siep, "Normative aspects of the human body," p. 174.

6. Robert Song, "The Human Genome Project as Soteriological Project," in *Brave New World? Theology, Ethics and the Human Genome*, Celia Deane-Drummond, ed. (London: T. & T. Clark, 2003), pp. 164–184 at p. 164.

7. Song, "The Human Genome Project," p. 174.

8. Mary Catherine Hilket, O.P., "*Imago Dei*: Does the symbol have a future?", *Santa Clara Lectures* 8:3 (2002) 9. Hilkert cites as an example a 1979 pastoral letter of the U.S. Catholic bishops ("Brothers and Sisters to Us"), which condemned racism because it violates the fundamental human dignity of those who, according to Genesis, were created in the image of God.

9. Paul Ramsey, *Fabricated Man: The Ethics of Genetic Control* (New Haven, CT: Yale University Press, 1970), p. 38.

10. Thomas A. Shannon and James J. Walter, *The New Genetic Medicine: Theological and Ethical Reflections* (Lanham, MD: Rowman and Littlefield, 2003), p. 8.

11. Shannon and Walter, *The New Genetic Medicine*, pp. 8–9. The stewardship model is more traditional while the co-creator model has been advanced by some contemporary theologians, especially Philip Hefner, Ted Peters, and Ronald Cole-Turner.

12. Douglas John Hall, *Imaging God: Dominion as Stewardship* (Grand Rapids, MI: Wm. B. Eerdmans, 1986), p. 115.

13. Hall, *Imaging God*, p. 107.

14. Claus Westermann, *Creation*, trans. John J. Scullion, S.J. (Philadelphia: Fortress Press, 1974), p. 56.

15. Thomas W. Mann, "'All the families of the earth': The theological unity of Genesis," *Interpretation* 45(4) (1991): 343.

16. Ruth Page, "The Human Genome and the Image of God," in Deane-Drummond, ed., *Brave New World?*, pp. 80, 84.

17. James F. Keenan, S.J., "'Whose perfection is it anyway?': A virtuous consideration of enhancement," *Christian Bioethics* 5:2 (1999): 104–120.

18. Julie Clague, “Beyond Beneficence: The Emergence of Genomorality and the Common Good,” in Deane-Drummond, ed., *Brave New World?*, pp. 212–213.

19. Heyd, “Human nature,” p. 165.

20. “Health is a state of complete physical, mental and social well-being and not merely the absence of disease or infirmity,” according to the preamble to the constitution of the World Health Organization as adopted by the International Health Conference, New York, June 19–22, 1946; signed on July 22, 1946 by the representatives of 61 states (Official Records of the World Health Organization, no. 2, p. 100) and entered into force on April 7, 1948. The definition has not been amended since 1948.

21. Allen Buchanan, Dan W. Brock, Norman Daniels, and Daniel Wikler, *From Chance to Choice: Genetics and Justice* (Cambridge: Cambridge University Press, 2000), p. 80. This work exhibits some of the internal tensions that one might expect of a multiauthored book. Elsewhere, it endorses what it calls a Rawlsian “political conception of justice,” the foundation of which is “intuitive ideas present in the political culture” (p. 147). In my view this is not an adequate foundation for justice or bioethics and is at odds with the idea that there are basic parameters of human functioning in light of which basic meanings of health and disease can be discerned, an idea reflected in the book’s reference to “restoration and maintenance of species-typical functioning” as a criterion for the public provision of genetic health services (p. 314).

22. Norman Daniels, *Justice and Justification: Reflective Equilibrium in Theory and Practice* (Cambridge: Cambridge University Press, 1996), p. 213, Italics in original.

23. Daniels, “Genes, Justice, and Human Nature”, in Buchanan et al., *From Chance to Choice*, p. 80.

24. Daniels, “Genes, Justice, and Human Nature”, in Buchanan et al., *From Chance to Choice*, p. 82.

25. Daniels, “Genes, Justice, and Human Nature”, in Buchanan et al., *From Chance to Choice*, p. 101.

26. Daniels, “Genes, Justice, and Human Nature”, in Buchanan et al., *From Chance to Choice*, pp. 96–97.

27. Daniels, “Genes, Justice, and Human Nature”, in Buchanan et al., *From Chance to Choice*, pp. 97–98.

28. Ronald Cole-Turner, “Do Means Matter?”, in Erik Parens, ed., *Enhancing Human Traits: Ethical and Social Implications* (Washington, DC: Georgetown University Press, 1998), p. 155.

29. Cole-Turner, “Do Means Matter?”, in Parens, ed., *Enhancing Human Traits*, pp. 157–158.

30. Erik Parens, “Is Better Always Good? The Enhancement Project,” in Parens, ed., *Enhancing Human Traits*, p. 14.

31. Eric T. Juengst, "What Does Enhancement Mean?", in Parens, ed., *Enhancing Human Traits*, p. 34.

32. Margaret Olivia Little, "Cosmetic Surgery, Suspect Norms, and the Ethics of Complicity," in Parens, ed., *Enhancing Human Traits*, pp. 162–176. See also Buchanan et al., *From Chance to Choice*, pp. 327–333.

33. Parens, "Is Better Always Good?", in Parens, ed., *Enhancing Human Traits*, p. 13.

34. Parens, "Is Better Always Good?", in Parens, ed., *Enhancing Human Traits*, pp. 13–14.

35. Thomas Szasz, *Pharmacracy: Medicine and Politics in America* (Westport, CT: Praeger, 2001), p. xvi.

36. U.S. government data cited by Cynthia Smith, Cathy Cowan, Art Sensenig, Aaron Catlin, and the Health Accounts Team, "Health spending growth slows in 2003," *Health Affairs* 24:1 (2005): 185–194.

37. Robert J. Barnet, "Ivan Illich and the nemesis of medicine," *Medicine, Health Care and Philosophy* 6 (2003): 276.

38. For information on the Global Fund, see its website, www.theglobalfund.org.

39. UN Millennium Project, "Investing in Development: A Practical Plan to Achieve the Millennium Development Goals," presented on January 17, 2005. Available at http://unmp.forumone.com.

40. Thomas H. Murray, "The Genome and Access to Health Care: Two Key Ethical Issues," in Thomas H. Murray, Mark A. Rothstein, and Robert F. Murray, Jr., eds., *The Human Genome Project and the Future of Health Care* (Bloomington: Indiana University Press, 1996), p. 215. Murray refers to another chapter in the same volume by Maxwell J. Mehlman, "Access to the Genome and Federal Entitlement Programs," pp. 113–132.

41. Marcia Angell, *The Truth about Drug Companies: How They Deceive Us and What to Do about It* (New York: Random House, 2004).

42. Angell, *Truth about Drug Companies*, pp. 91–92.

43. Angell, *Truth about Drug Companies*, p. 11.

44. Angell, *Truth about Drug Companies*, p. 200.

45. Angell, *Truth about Drug Companies*, pp. 193–196.

46. Angell, *Truth about Drug Companies*, p. 102.

47. Angell, *Truth about Drug Companies*, pp. 101–104.

48. Sheldon Krimsky, *Science in the Private Interest: Has the Lure of Profits Corrupted Biomedical Research?* (Lanham, MD: Rowman & Littlefied, 2003), p. 9.

49. Krimsky, *Science in the Private Interest*, pp. 178–179.

50. Mark S. Frankel, "Inheritable genetic modification and a brave new world; Did Huxley have it wrong?", in *Hastings Center Report* 33: 2 (2003) 32.

51. Frankel, "Inheritable genetic modification, p. 36.

52. Buchanan et al. *From Chance to Choice*, p. 340.

53. Angell, *Truth about the Drug Companies*, p. 259.

54. See, for example, the websites of the Catholic Health Association (www.chausa.org), the Carter Center in Atlanta (www.cartercenter.org), and the Center for Genetics and Society (www.genetics-and-society.org: index.asp).

55. For more on this topic, see Lisa Sowle Cahill, "Bioethics, theology and social change," *Journal of Religious Ethics* 31: 3 (2003) 363–398.

# 8

# Freedom, Conscience, and Virtue: Theological Perspectives on the Ethics of Inherited Genetic Modification

Celia Deane-Drummond

There can be little doubt as to the real possibility of inherited genetic modification[1] in the coming years. Such possibility brings new knowledge but also the possibility of a redefinition of what it means to be human—human identity as such. While the ontological existence of a clearly defined human nature can be challenged both philosophically and scientifically, the term "human identity" in this context relates more to human meaning and purpose rather than essentialist concepts about human nature as such. The possibility of IGM raises important questions about religious meaning; for example, do such reformulated identities at the borderline of natural and artificial threaten religious understanding of what it is to be human? In other words, are we approaching what some have described as a posthuman condition?[2] While some artificial intervention in human life is inevitable in modern medicine, IGM seems to go further in inaugurating the possibility of more permanent changes that are passed on to subsequent generations. In addition, how far can such changes be recognized as licit in terms of theological anthropology?

Rather than explore all the different facets of what has variously been termed theologically as humanity made in the image of God, I intend to highlight in this chapter those features of the human person that are particularly relevant to the discussion of IGM in terms of ethical practice. In particular, I explore the question of what does it mean to have freedom in the context of new scientific knowledge and application of human genetics, which itself may undermine more traditional notions of the human. Hence does IGM appear to advocate freedom for scientists but lead indirectly to a loss of freedom, rather than its gain, for those individuals who are the designed "products" of such technology? What is the role of individual and collective conscience and what does it mean to act as a scientist true to conscience in such circumstances? Finally, how might a Christian understanding of virtues, including wisdom, prudence (practical wisdom), humility, and justice, situated in the context of the theological virtues of faith, hope, and charity, serve to qualify the goals and aims of IGM?

## The Relationship between Freedom and Knowledge in Genesis

Freedom has become a byword in western contemporary society, the notion of liberty intricately bound up with our sense of self and who we are as persons. Human autonomy and the giving of free consent constitute core values in which much bioethical reflection and clinical practice seem to be rooted. However, does this mean that all scientific endeavors should be permissible, based on the notion of freedom of the individual as long as consent is also free? The fact that some scientific activity is either legally controlled or outlawed suggests that notions of the common good qualify individual freedom. The common good in turn is viewed differently in different societies, as exemplified by the different legal restrictions on human cloning in different parts of the world.[3] The high place given to human liberty in countries such as the United States may be one reason why there are relatively few legal restrictions on genetic experimentation, although of course in the United States in particular, the politically explosive nature of debates about the status of the human embryo tends to freeze political discussion and legislation.[4] In order to begin to answer questions about freedom of the individual in the context of human genetics, especially in moving to the uncharted territory of IGM, we need to consider the prior question: What is the relationship between freedom and knowledge? Is all knowledge, including knowledge of human genetics, and more specifically its practical application, necessarily good from a theological point of view?

Some hints at the way this question might be resolved from the perspective of Christian theology appear in the first book of the Bible, namely the Genesis text. Genesis 2: 16–17 begins with a positive command given by God to humanity: "You may eat all the fruit of all trees." But then there is an exception, the tree "of the knowledge of good and evil you may not eat." Why not? The reason given in the text is simple: The consequences are dire; it leads to death. This is, of course, the account of the fall of humanity, following which humanity was expelled from the Garden of Eden. But why did God give this command? Is God suggesting that some knowledge is impermissible? Biblical scholars divide over this issue. Some, like von Rad, say that we can never know why God gave this command and it is even illegitimate to ask.[5] Claus Westermann agrees and elaborates further that the insertion of "knowledge of good and evil" as a description of the tree was added late in the construction of the text, hence it was originally understood simply as the tree in the middle of the garden.[6] So it is misunderstanding the text to ask why God forbade such knowledge. Moreover, the structure of the passage as command and consequence is as a directly spoken word from God to humanity, a clear signal that this story belongs to primeval time rather than historical time.

The interpretation of Genesis 2 can be viewed as indicating the possible freedom in which humans now operate—that they now have the ability to choose yes or no. This is a newly acquired freedom of choice that was not there previously. However, to say no to God is to say no to life, hence the consequence—death. Like other religious taboos, such as rules for eating in early Israel, there is no rational basis given in this case for such a prohibition; it simply was there as setting a boundary condition. It is important to note, with Westermann, that "where human freedom means utter lack of restraint and arbitrariness, then human community and relationship with God are no longer possible."[7]

The couple in the garden felt no shame before eating the fruit of the forbidden tree, but prompted by the serpent they ate, and the consequence was that they became full of shame. Other religious traditions spoke of a higher magical knowledge that could be gained by means of a serpentlike creature, or alternatively the serpent represented a fertility cult, encouraging sexual relations. Does the story simply warn against such religious cults? This is unlikely because the tale also speaks of the serpent as one who is made by God, who through its cleverness led humans astray. Why does the snake act in this way if God made it? The answer never comes; rather, the origin of evil in this case is not known or clearly understood. The author, then, is aware of the riddle and mystery of evil in the midst of a creation declared by God as good. As if to reinforce this riddle, the knowledge itself that the snake proffered was partly correct, it did lead to a different sense of self in relation to others. In this sense one could argue that the Fall is not wholly negative because it led to human maturing that would not otherwise have been possible. Perhaps we should note that the knowledge of good and evil spoken of by the serpent and throughout the Genesis account means good and bad in a general sense, not good and evil in a moral sense. Such knowledge is always ambivalent; it can elevate life or put it in danger. Much the same could be said about knowledge acquired through the study of human genetics, quite apart from its application.

Ezekiel 28: 11–19 also speaks of primeval quasi-divine persons acquiring divine wisdom and then being sent from a mountain; in this case it was the sin of pride that led to expulsion. Other religious traditions spoke of human aspirations to life and knowledge. While life could be cut short by death, there seemed to be no limit to knowledge. The temptation story in Genesis is different in that the actors are not semi-divine, nor do they acquire divine wisdom; their attempts to cover their shame using fig leaves apparently failed, and God intervened to give them skins to cover themselves. It would be a mistake to read into the story specific developmental stages toward civilization; rather, it is primarily a story about human fallibility.[8] The temptation account is perfectly understandable and natural from the perspective of the actors in the story. New possibilities seem to be opened up by the snake's offer; Adam merely conformed to the trend and thereby committed a passive form of sin.

This is the nature of all temptation, and the account of the fall of humanity is archetypal in this respect. Temptation appears in the guise of the good, but is more fundamentally not so; however this is only apparent in retrospect. Mutual support in error is a counterpart to the positive mutual support in community. The outcome of the transgression is ambivalent; while innocence is lost, the knowing is partly positive—there is a shift in sense of consciousness, of what it is to be human.[9] The story is clear that the couple did not initially feel any sense of guilt in what they had done; they had to be told of the crime before they realized it was wrong and only then did they attempt to defend their action. In their defense they turned away from God. This universal sense of alienation from God applies to all people, not just the Israelite nation.

From this account we can conclude that according to the Genesis text, human nature is distinct from the animals in its ability to choose. The outcome of such choosing may be partly positive, but such ability goes hand in hand with fallibility to break God's commands without even necessarily being aware that this is the case. Freedom is not arbitrarily choosing whatever is possible but needs to be situated in the context of relationship with God and others. We can conclude that there is no warrant for assuming that choice in itself is detrimental to human life. Rather, it is the way choices are made that is important; in other words the goals to which those choices are directed, either in alignment with or against what is understood as God's goals for humanity. Freedom is positive and in one sense the serpent is right: Choosing in a self-conscious way is a new departure for humanity, a higher plane of thinking "like one of the gods." If the Fall is about turning away from God's command, then redemption similarly implies living in alignment with those commands. How are we to understand from a theological point of view what it means to have freedom? In what sense might there be any limits to the scope of human searching after knowledge, particularly in the context of genetic knowledge?

Rahner describes knowledge as one act of freedom, and the extent and plurality of knowledge amounts to concupiscence, as inordinant desire; hence theological knowledge and scientific knowledge need to exist alongside each other, without one taking over the other. As such he suggests that they "inevitably threaten and disturb each other."[10] For Rahner, theology unmasks those sciences that make claims beyond themselves in a dominating fashion, while at the same time theology itself is challenged to reexamine itself. Negotiating the way genetic science and theology threaten but challenge the other to reconsider its own perspectives on the meaning of freedom is a task of some complexity. In order to develop this discussion I will argue that freedom is an integral aspect of our humanity, our human dignity. However, it is equally important to develop theological notions of freedom that are robust enough to withstand secular versions that may or may not cohere with Christian theology. In addition, the way freedom is understood in theological terms puts a fresh light on more ethical aspects of personhood, such as conscience and the virtues.

## *Imago Dei* as Freedom

The concept of freedom is at the heart of our existence and core of our experience, but what does it mean? Is there a sense in which freedom might be redeemed from the negative consequences of the fall of humanity? In addition, given the fallibility of humanity, how might we understand freedom in such a way that it reflects who we are in relationship to God, rather than fostering a breakdown of that relationship? In other words, can human freedom be considered an aspect of what it means to be in the image of God while recognizing the temptation to put the human self in the place of God in expressing what that freedom might mean?

Popular and modern understanding of freedom has tended to follow what is broadly called the freedom of indifference, which subsequently leads to theories of moral obligation. Freedom of indifference originated in the fourteenth century in the work of William of Ockham, a Franciscan who worked out a new concept of freedom in association with nominalist philosophy. While the scholastics argued that freedom came from a prior sense of reason and will, Ockham believed that freedom was primary and that intelligence and will presupposed free will. For Ockham, freedom consisted in the power to say yes or no, to be totally indifferent with respect to the whole range of possibilities set before one. Natural willing was also indifferent, so a pure will had to be imposed from within oneself, as it were, in order for an action to be right. The contrast with the Thomistic view could not be clearer. While for Thomas free will was directed toward the good through natural inclinations, for Ockham natural inclinations had to be opposed because they reduced the scope of freedom. The moral outcome of such a theory is that individual conscience is a primary factor in controlling actions, but it is a conscience aligned with an apparently changing and somewhat arbitrary will of God. The theologian Servais Pinckaers suggests that "Beneath freedom of indifference lay hidden a primitive passion—we dare not call it natural: the human will to self-affirmation, to the assertion of radical difference between itself and all else that existed."[11]

In other words, freedom consisted in being able to choose from as wide a range of possibilities, including capricious alternatives. Loyalty became a threat, for it limited freedom. Reason as such had no hold on freedom, and the will had to be ordered according to certain laws, by commands and through obedience. Only the will could direct right ordering according to the commandments. God's free will became the focus for understanding who God is, rather than an emphasis on God's love, wisdom, and truth as in the earlier classical understanding. In this scenario, humanity was held under the obligation of God's law. Hence it experienced God's law as restriction and limitation. Knowledge that was impermissible was, in theory, knowledge outside the moral law of God given to humanity. Yet because God was absolute, he was beyond such moral laws, so God could command otherwise, leading to a relativist view where the only stipulation was the certainty in conforming to God's will. God's will was first that found in the Scriptures, which expressed universal and stable precepts, and second that revealed through individual conscience, but such conscience was in conformity with the revealed will of God. God could command all manner of different possibilities, so all manner of genetic interventions could be justified by this understanding of freedom, as long as the individual believed that this was God's will. Bizarrely perhaps, if ethical precepts were common to all people, then they could exist independently of God, so that moral theory could be severed from any notions of God and worked out simply through individual conscience. The following tensions arise through such an understanding of freedom:[12]

1. Freedom or law.
2. Freedom or reason (reason restricted freedom; freedom could be irrational).
3. Freedom or nature (nature itself would lead to limitation, so freedom had to dominate nature as well).
4. Freedom or sensibility and emotions. Freedom may be aligned with emotions or against emotions.
5. Freedom or grace.

6. Freedom of man or God; either was exalted but not both together.

7. Subject or object. Science's greatest danger was viewed as allowing subjectivity to creep in; for the humanities, it was treating a person as an "object."

8. Freedom of self or others. The freedom of others felt like a limitation on individual freedom. This is related to:

9. Freedom of the individual and society.

10. Illicit knowledge as outside individual conscience or "divine"/legal command.

For freedom of indifference, the ability to make decisions is independent of motives. A popular contemporary view of freedom is a modification of this view, namely the ability for self-determination through acknowledgment of motives. However, underlying such a view is the assumption that freedom means the power to choose between different sorts of action.

What happens if freedom is totally open ended? We arrive at an anarchic form of freedom that is haphazard in shape and form. It may lead to partial goods, in the manner of the Genesis story, but it is not directed to any goal other than the limited goods of human desire, which may or may not be misdirected. Might there be an alternative understanding of freedom that is more appropriately related to humanity as made in the image of God? Those who have learned a skill or craft, or have a musical talent know that once such an art is mastered, it leads to a new kind of freedom, one that we could call the freedom for excellence.[13] While the freedom of indifference opposes natural inclinations, freedom for excellence uses them. Virtues are not rejected as restricting freedom, but are the means through which freedom develops. Initially it requires discipline, but the ultimate intention is to expand freedom, leading to flourishing by removing excesses.

For Thomas Aquinas, the first stage in the development of freedom is the formation of charity, love of neighbor and love of God.[14] The second stage moves beyond commandment expressed as punishment or reward and positively toward the practice of virtues, including the three theological virtues of faith, hope, and charity, drawing on the teaching of the Sermon on the Mount. The third stage is a response to a call, a vocation, where free actions are like the fruits of a tree arising out of a center of freedom. The final stage of perfection flowers from desire for unity with God, acting within the grace of the Holy Spirit. Such freedom allows an integration of the polar opposites that unfold in the freedom of indifference, so there is a working with nature, a union of grace and freedom, subject and object, freedom and law, freedom and reason, self and others and so on. Knowledge is no longer hedged about in a rigid way through law and conscience as in a freedom of indifference. The possibility of mistaken knowledge is always there like a shadow because of human fallibility, as the Genesis story reminds us. However, in the context of freedom for excellence, the likelihood of excess is reduced; the focus is no longer negatively on what I should not do, but on what I can do, in a positive way in order to express virtues.

How far can such claims be recovered in a contemporary context? Historical sensitivity points to the real restrictions on human freedom that arise out of social, cultural, and political restraints. The postmodern context of contemporary western culture is also a sharp reminder of the historical situatedness of any understanding of human identity, including human freedom. The theologian Karl Rahner has pointed to the importance, not just of a historical appreciation of our limitations in a discussion of freedom, but also a venturing, a looking forward to the future so that we can see more clearly what we might become.[15] He also, significantly, moved away from an understanding of human nature in terms of essences, which is presupposed in Thomistic theology, and toward one that puts much more emphasis on existential experience. However, he still affirms the idea that freedom is a defining characteristic of human dignity and that this freedom needs to be understood in terms of relationality. In particular, the fundamental option or choice is toward good or evil, echoing the Genesis text.

For Rahner, true freedom is basic and fundamental to human personhood and it arises out of living a life in orientation toward the good, reinforcing the kind of freedom articulated in Thomistic thought. He calls this form of freedom transcendental freedom; it is an acceptance or rejection of a loving relationship with God.[16] It is this transcendental freedom that is the defining characteristic of persons, for through it the moral character of the person grows and develops, making choices and decisions. These day-to-day decisions are the outcome of what Rahner terms categorical freedom. It is the concrete expression of the fundamental option to orient oneself toward God rather than against God. Above all, Rahner argues that freedom of the person grows as she or he becomes more dependent on God, so that "it is responsible self-mastery, even in the face of God, because dependence on God—contrary to what takes place in intra-mundane causality—actually means being endowed with free selfhood."[17] Once freedom is viewed as the deepest response of humanity in relation to the freely given love of God, it becomes an essential ingredient in the dignity of the person. It is this foundation that leads, thereby, to an understanding of what the freedom of the conscience means in practice, freedom of choice and Christian freedom standing in the same relationship as nature and grace.

Does this mean that those who are unbelievers cannot enter into transcendental freedom? This question is partially resolved by Aquinas' notion of natural law theory, which is a fundamental decision to do good and avoid evil that he believes is common to all humanity, whether or not they recognize God. In addition, a full answer to the question needs to take account of the way freedom is expressed through the virtues; hence those who develop those virtues are traveling on the same path, whether from an unbelieving or believing perspective. This is implicit in Rahner's understanding of the relationship between nature and grace, freedom of choice deriving from nature and transcendental freedom arising from grace. Rahner summarizes a theological understanding of freedom thus: "[It] is self-mastery bestowed on man in the dialogue with God, where he is called to the finality of love's decision."[18] Of course, this also draws on the classic tradition since Aquinas recognized that virtues may be both learned and/or acquired as gift from God. Rahner was reacting to the legalistic approaches that evolved in neoscholastic thought, which downplayed both the importance of the individual and historical contingency. It is important to insist that the workings of the Holy Spirit are never restricted to those who are believers.[19] However, at the same time, one might anticipate that the most developed sense of virtue will be formed in the context of the experience of religious communities. Certainly a sense of obligation and deliberate option for good is likely to be more explicit in those who have a religious belief. As such, a Christian perspective on issues, whether as a practicing scientist or not, can act as a guide to certain ways of orienting one's self in relation to others, but this is not intended to exclude those who come to similar conclusions through different means.

Is such a view of transcendental freedom too idealistic? Any idealism needs to be countered by the memory of the fall of humanity with which this discussion about freedom began. According to this account, freedom is necessarily ambiguous although this ambiguity was rooted in the concept of freedom of choice, rather than freedom in relationship with God. Rahner was also acutely aware of the temptation toward utopias in considering absolute freedom and actual opportunities as they arise.[20] He argued that Christianity has a built-in claim against any absolute or infinite freedom as a real possibility, for it "deprives man of his illusion that infinite freedom can be attained in the course of history itself."[21] The very act of struggling to attain freedom produces new limitations. He also explored attempts to attain more human freedom through genetic manipulation, alongside sociological or psychological manipulation. Yet he believed that all such attempts inevitably start from limited facts that then serve to restrict the other's freedom.

What is the practical outworking of such a view in relation to IGM? I suggest that such a view implies looking beyond the immediate prospects to the longer-term consequences, to see how such a future coheres with the good of the human race, including wider social issues such as those involving political justice. Few could doubt that human identity is grounded in freedom. I have also argued that understanding freedom as rooted in an overall choice to work for good and not for evil serves to shape the way freedom is expressed while not hedging its limits in a legalistic way. Manipulation of the self, including genetic manipulation, is neither hell on earth nor the coming of God's kingdom, but is situated in a world where it is possible for humanity to work for the development of its own freedom in the sense that humanity is always existentially both radically open and incomplete.[22] However, concrete realization of manipulation of a human being in a way that can be planned, controlled, and regulated is a new departure for humanity; hence it presents new challenges. Human freedom implies the possibility of such manipulation, so that IGM does not necessarily automatically imply a morally repugnant act.[23] Its moral acceptability depends on a number of factors, such as the source of the manipulation, its intention, and the means used to achieve particular ends. I will discuss practical moral aspects later; the intention here is to explore how far IGM is in principle a genuine act of freedom that fosters our humanity.

Genetic knowledge is always ambivalent because it can be used in particular ways to affirm or deny qualities of human persons. I have my doubts that the kind of scenario envisioned by Francis Fukuyama, namely an inevitable dehumanizing effect of genetics on the human race as such, is at all realistic. It makes for entertaining reading and discussion, but is it true to the scientific possibilities currently in place? Fukuyama assumes that genetic manipulation will become commonplace and routine rather than limited in scope. It is important to remind ourselves of the fallibility of human nature, but in this case such fallibility has also been expressed by an overoptimistic estimate of what might be possible; one might even say it is the sin of pride in the possibilities open to humanity. It is easy to see why a logical consequence of such an extensive application of IGM might lead to the kind of dehumanizing scenario he is envisioning. However, this is based on a largely false optimism about the scope of IGM and its applications. Ted Peters has reminded us we are far more than just our genes.[24] The regulation of gene expression is also far too complex to contemplate what might happen if attempts were made to modify more complex genes involved in human behavior. Nonetheless, an acceptance of IGM as routine could be envisaged as a retrograde step. Instead, I suggest that the decision-making process needs to be broad enough to take into account long-term factors while narrow enough to consider cases on their own merit.

The difficulty with a case-by-case approach used in exclusion of wider political factors is that the broader issues are no longer aired. On the other hand, if broad trends are simply extrapolated from the initial optimism that surrounds new technologies, then this is also likely to lead to false conclusions. A full working out of what might or might not be permissible in particular cases of IGM enters the realm, not so much of freedom, but of conscience and practical wisdom or prudence. In answering the question of whether or not IGM will serve God, questions about how far this represents an expression of virtue are important because they take the spotlight away from the immediate dilemmas associated with each case and move it to wider issues about who we are becoming in practicing such technologies. In other words, how far is such freedom an expression of goodness? On the other hand, the virtue of prudence is particularly concerned with making correct judgments and thus is directly relevant to problematic cases. Before turning to discuss prudence, it is also important to consider first the notion of conscience, since medical ethics has tended to stress personal autonomy to such an extent that the individual conscience becomes the arbiter of moral judgments.[25] Furthermore, it is conscience understood in a particular way in relation to freedom, which in turn is understood as unrestricted choice.

**Reclaiming the Classic Notion of Conscience**

It is important to stress that even in freedom for excellence, or as Rahner prefers, transcendental freedom, there is still a place for law and conscience, worked out through categorical freedom.[26] It is also fair to say that daily decision making and adopting habits of virtue reinforce freedom for excellence or transcendental freedom. Thus there is dialectic between both forms of freedom rather than a simple application of the fundamental option.[27] For Aquinas, natural law is common to all and is the framework in which virtues develop. Even natural law showed—through a natural habit known as *synderesis*, the basic knowledge of good and evil—that good is to be welcomed and evil avoided.[28] The capacity for truth and goodness is the essence of freedom. If evil is done, it is due to a lack of freedom, rather than because of freedom. The will becomes oriented toward love rather than being used to dominate the self or others. The knowledge of good is the foundation of freedom. Knowledge of evil is, by contrast, a profound lack of freedom. Hence we might say that the wrong kind of knowledge is knowledge that is not in accordance with the freedom for excellence. In practice, owing to the human fallibility described so clearly in the Genesis account, perfect knowledge of the good can never be attained because it is rooted in a painful demand, so that it never perfectly coincides with our own idea of happiness. The Genesis story is a permanent reminder of the fallibility of human nature in spite of a vocation to the good. Aquinas wanted to insist that while the good was capable of winning the desire of all, he was quite aware of the inevitable failures to reach that end.

What might be the place of human conscience in debates about human identity? It might seem at first glance that conscience has a place in freedom understood as a freedom of indifference rather than freedom of excellence. Certainly, as Linda Hogan has pointed out, conscience has become the means through which human persons understand what it means to be morally good.[29] In the biblical tradition, conscience is normally associated with having integrity, where a pure heart, a good conscience, and a genuine faith are bound up with one another.[30] But what happens if an individual acting out of conscience opposes the teaching of the church? The conflict between conformity to church teaching and individual acts of conscience is highlighted by historically diverse interpretations of the theological meaning of conscience. I have suggested so far that freedom for excellence is a useful paradigm for understanding the meaning of human freedom. How is conscience understood according to writers in this vein?

As mentioned earlier, Aquinas believed that *synderesis* is the habit of practical reason arising out of natural law, the first principle of which is to do good and avoid evil. Conscience refers to the way those principles are specifically applied in individual circumstances. However, because the rule of *synderesis* is so general, to see conscience as a mere application of rules is to miss the point; rather, conscience takes a range of factors into account before reaching a particular judgment. For Augustine, conscience is never binding where it contradicts God's law. Aquinas resisted this explanation since he believed that if natural law applies—that is, that humanity is oriented toward the good and avoids evil—then to imply that conscience will knowingly contradict God's law makes no sense; rather, it will be viewed as a decision that leads to the good. Furthermore, a law cannot bind a person who is incapable of knowing its precepts. In other words, the conscience follows the good as perceived good, so to act against this is to act against reason, which is impermissible.

There are, of course, situations where the conscience is "erroneous." In this case a person is culpable if they were capable of knowing morally relevant circumstances and these were ignored. Brian Davies takes the view that for Aquinas, conscience is not specifically a moral virtue; rather, it is simply the judgment that we make about the goodness or evil of a particular moral action.[31] This interpretation differs from that if Linda Hogan, who seems to be more inclined to include action under the category of conscience, to see action as integral to the workings of conscience, and hence conscience as a moral category.[32] Certainly if we go back to Aquinas' texts on conscience, it is clear that he suggests conscience is an application of knowledge to something, which itself is an act.[33] He describes three ways such application can apply to conscience. The first is as recognition that we have or have not done something; in this sense it acts as a witness. The second is as recognition that we should do something; in this sense the conscience is said to incite, or bind (*conscientia antecedens*). The third is as a judgment of whether or not something has been done well (*conscientia consequens*). It seems, then, that action in the meaning that Aquinas portrays here is in an internal sense an act of judgment about oneself. In this sense Davies is correct to say that conscience is morally neutral. However, Hogan also points to the more practical consequences of what to do with the "whispering" of conscience. While in a technical sense conscience does not direct action, in a practical sense action follows the lead of conscience in that a conscience will witness, incite, or confirm particular actions. Aquinas also suggests that where a conscience is in error, "every act of will against reason, whether in the right or in the wrong, is always bad."[34] He even goes further than this in suggesting that if the reason believes this to be the will of God, then slighting this amounts to slighting God's law. Thus, as indicated earlier, he does not pitch the divine law against conscience in an Augustinian manner; rather, law and conscience work together. Moreover, a person cannot be held culpable for those errors of conscience that arise out of unavoidable ignorance, only those that arise out of avoidable ignorance.

Could scientists who sincerely believe that they are fostering the good in following their own conscience with respect to IGM be judged as acting immorally? According to Aquinas, as long as a scientist sincerely believes that he or she is acting in good faith and for the wider good of the community, their views need to be respected. Furthermore, it would be wrong for them not to take notice of their conscience and follow its lead. However, the example suggested here shows the limitations of arriving at ethical judgments through considerations of conscience alone. It is all too easy to translate our modern understanding of conscience into Aquinas' apparently liberal view without taking sufficient account of the severe restrictions in which his own society functioned. Certainly, individual human freedom and autonomy were not valued as highly then as they are today. Thus Aquinas' view could be seen as a refreshing counterbalance to an overriding trend toward hegemony. In addition, his view of conscience was always optimistic about human nature—that *synderesis* applies to all human persons and that we are naturally disposed to follow good and avoid evil. This optimistic view of human nature needs rather more qualification in light of our more contemporary understanding of the ability for self-deception, although of course he did try to take this into account. Finally, having a clear conscience is just one factor to consider. I argue later that conscience needs to be situated much more strongly in prudential decision making for it to be relevant for theological ethics.

It is important also to stress that for Aquinas conscience is not simply a bald act of reason but is a combination of two prior traditions of conscience, one stemming from Bonaventure that stressed the importance of the will and one stemming from Albert that stressed the importance of the intellect.[35] It is one reason, therefore, why Hogan believes she is justified by tradition in arriving at a personalist view of conscience, which brings together the reasoning, emotional, and spiritual aspects of personhood. While such an enlarged view of conscience is a helpful counter to more narrowly restricted views, I suggest that in some respects conscience is being asked to carry elements that would more naturally fall outside its capacity as conscience. Of course in many respects the classical view of conscience was naïve in relation to our current understanding of human psychology and sociology of knowing. However, if conscience is situated more clearly in the tradition of virtue ethics as an element of prudence, then the difficulties encountered in fitting together a more holistic understanding of persons and conscience no longer apply because prudence or practical wisdom is an intellectual virtue of practical reason that includes judgment but also leads to action.[36] Prudence is also something that can be developed and enlarged to include the wider human community. While it is possible to enlarge notions of conscience to do the same by situating individual judgments in wider social contexts, I would argue that it is more difficult to do so, for in the last analysis conscience is about my own judgment about my actions, whether or not I take into account other human relationships. Prudence also needs to be situated in a broader context of what it means to have particular virtues.

## Uncovering a Virtue Ethic for Inherited Genetic Modification

Which particular virtues are most relevant in considering new knowledge? Aquinas allowed both wisdom, directed toward uncovering eternal truths, and knowledge, directed toward uncovering contingent truths. Both wisdom and knowledge are relevant sources for moral action that need to be considered. There was no sense in which Aquinas believed that some areas of knowledge were impermissible, but instead that information from both wisdom and knowledge can be relevant in a practical sense in deciding what to do according to practical reason. Thus it is one matter to know what it is possible to do through IGM, it is quite another matter to apply this knowledge to particular circumstances in circumscribed ways. I have suggested so far that the context in which we should think about new knowledge is that of freedom understood in terms of excellence, including a transcendental view of freedom, as Rahner would have indicated.

For a Christian, other virtues are significant as well. Aquinas describes the seven gifts of the Holy Spirit as wisdom, understanding, counsel, fortitude, knowledge, piety, and fear of the Lord.[37] The first gift, wisdom, is of particular significance. In as much as wisdom can be learned, it is a virtue that can be shared by all those of good will, whether or not they are Christians. In this sense it is aligned with the idea of natural law. Divine wisdom also finds expression in the eternal or divine law, which for Christians is expressed in Christian discipleship (Ephesians 1. 8–10). Therefore a measure of whether an action is wise is its relationship to this divine law, a point made repeatedly by Aquinas in his *Summa Theologiae*, especially in the third part.[38] As I suggested earlier, freedom for excellence seeks to move beyond the restriction imposed by law by acting according to the gift of the Spirit.

For Aquinas, wisdom is one of the three intellectual virtues of speculative reason, the others being understanding, or grasping first principles, and *scientia*, which denotes the comprehension of the causes of things and the relationship among them. In other words, wisdom is the understanding of the fundamental causes of everything and their relationship to everything else. Human wisdom is a virtue directed toward the wisdom of God, for while wisdom can be learned, it cannot be grasped or used for human aggrandisement (Proverbs 16). In the fullest sense, human wisdom is possible only through the gift of the Holy Spirit by the grace of God. The Christian vocation includes developing the virtue of wisdom.

Wisdom is closely related to one of the intellectual virtues of practical reason, namely prudence, also termed practical wisdom. Practical wisdom is particularly significant for ethics because it sets the way individual virtues must be expressed in particular circumstances, or broadly speaking, the means of attaining a virtue. Developing prudence is not just about one's inner attitude, although it includes this, it is also about how this attitude is expressed in action. The three elements of reasoning integral to prudential decisions are taking counsel or deliberating, judging, and acting. Aquinas is insistent that the goal of the action needs to be recognized clearly. This may involve taking counsel, which would include consulting a wide range of opinions. What are the aims of IGM? Is it to permanently remove a particular genetic predisposition to a given disease, in which case those affected by that disease would need to be consulted? Is it to enhance profits for a company? While this may be an indirect consequence of all commercial medicine, if it is a primary goal, it needs to be evaluated according to who is going to benefit the most and whether it can be justified. What are the wider consequences in terms of social effects in channeling resources to some goals rather than others? What are the long-term effects on the human population, if any? What are the means that are going to be used to effect these changes; for example, will it involve destruction of embryos or human cloning?

Practical wisdom comes into play in discerning the most appropriate way of acting in given circumstances, as well as the goals of such action. If either the goal or the means are faulty, then this leads to sham prudence. A choice that is made about the most appropriate means of acting in a prudential decision is what Aquinas would term the act of conscience.[39] Forms of discernment that act against the needs of the community amount to folly. Perhaps more accurately we could say that the goal is a partial good, for it benefits relatively few people. A virtue ethic includes the idea of consequences but is oriented toward the common good. Aquinas also used the term "incomplete" prudence to indicate when the good is narrowed to particular individuals. In this case it becomes obvious that whereas conscience is important, it is situated in an overall prudential decision-making process that qualifies its significance. Daniel Westberg has put this more strongly: "The equation of prudence with conscience is still faulty: conscience becomes the voice of reason, and the role of prudence is reduced to the perfection of the judgment of conscience. This does not necessarily result in good actions if the agent's will is contrary . . . The agent may not actually follow his conscience, and so not carry out his best judgment."[40]

Thus, it is far too limiting to reduce prudence to conscience. Even if conscience is enlarged, it does not follow through with the moral act that is an integral aspect of prudential thinking. In addition, once conscience is situated in the wider context of prudence, rather than the other way around, it becomes much more obvious where possible difficulties may be encountered and this leads to a more realistic understanding of human nature than might be possible if conscience is abstracted from the rest of practical reasoning. Thomas was insistent that the primary clause, that good is sought and evil avoided through the principle of *synderesis*, could not be mistaken; it acted as a guide for the rest of his thinking.

In addition to *synderesis*, morally relevant principles can come from *sapientia* (wisdom), which discerns divine obligations, and *scientia* (knowledge), which discerns knowledge about the natural world. Both of these areas could be mistaken; a scientist might decide that there is no evidence that a particular action carries a risk, but when a new situation arises, this conclusion may be in error. This is particularly significant in the context of IGM because many of the possible experiments cannot be performed easily without error. Peter Vardy believes that some degree of acceptance of mistakes is inevitable for scientific progress.[41]

The question that is ripe for discussion is how many mistakes can be tolerated, rather than imply, as he does, that scientists will know as a matter of course when or when not to take those risks. In addition, when it comes to consideration of IGM, the tolerance of a mistake needs to take into account the seriousness of such consequences for both the individual human life and the family and society in which that life is lived.[42]

Of course I am not implying by these remarks that scientists are always irresponsible or that scientists and ethicists are pitted against one another in discerning moral boundaries for action. Nothing could be further from the truth. Rather, in matters of such fundamental importance as IGM, a degree of consensus needs to be reached by a community as to whether a particular action is justifiable and how far it is justifiable. Christians entering into the debate will also want to know how far such actions are in alignment with their understanding of divine obligations. At the same time, it is entirely possible that just as scientists may be mistaken in their understanding of genetic science, so Christians may be equally mistaken in their understanding of divine obligations. The variety of interpretations of what is legitimate from a Christian perspective only serve to emphasize this point. Even while the aim will be to serve divine law, such an understanding of divine law should not be thought of as legalistic or alternatively, arbitrary or changeable, which tended to be the tradition arising out of the Ockham school. Rather, the principle of goodness needs to be situated in a theological context where charity in particular has a role to play in judging whether to act in particular circumstances.

Following Aquinas we can also suggest that there are ways and means of reducing the likelihood of arriving at an incorrect judgment, or erroneous conscience, alongside a distorted prudence. In the first place there is no excuse for ignorance, that is, ignorance of how IGM can take place, at least in general terms, be it through modification of individual gametes or embryos, or through some form of nuclear transfer or cloning technology. However, Aquinas would oblige someone to follow their conscience in making prudential decisions in those situations where an authority is forcing a situation perceived to be unjust.

What would count as behaving virtuously in the context of IGM? Practical wisdom includes a number of characteristics that are worth pondering in the present context. Aquinas draws on five areas related to knowing; namely, memory, reason, understanding, aptness to being taught, and ingenuity.[43] While Aquinas takes this list from Aristotle, biblical wisdom also includes similar ideas of memory and aptness to being taught (Proverbs 3.1), reason and understanding (Proverbs 2.5, 18.15), and ingenuity (Proverbs 8.30). Clearly, reason, understanding, and ingenuity all come into play in developing science and expanding the horizons of knowledge, but these characteristics alone are not sufficient for practical wisdom since it includes memory as well.

### Inherited Genetic Modification: A Critical Appraisal

When Paul Ramsey first raised objections to IGM, he did so in the context in which geneticists freely talked about the possibility of using genetics as a eugenic means of controlling human evolution, of removing those genes thought to be deleterious to the human race as a whole.[44] Scientists today are far more aware of the political explosiveness of such remarks and are more measured in their arguments for the limited usefulness of genetic engineering based on specific medical needs, rather than any aspiration for population changes. Ramsey argued that in principle there can be no moral objections to IGM as such where this is to remove a deleterious gene, assuming that the risks associated with the process have been eliminated.[45] Certainly it would seem to be expressive of charity in action, where charity is aligned with those parents who are desperate to have a child of their own free of genetic disease. Ethical questions do arise, however, in relation to the means used for such a change and the possibility of limited knowledge of the complexity of the process, leading to unforeseen effects. Yet, even where all knowledge is taken into account, Ramsey suggests that some errors may be made, but this is inculpable rather than culpable ignorance.[46] He suggests that waiting until all errors are impossible would be an unreasonable moral demand. Such positive appraisal of genetic engineering counters his more strident comments about genetic engineers playing God, although his critical remarks were directed toward a theological blessing of all that is possible.[47]

Ramsey also suggested that Rahner's approval of manipulation of the individual amounted to a carte blanche for all genetic technologies.[48] However, he seems to have misunderstood Rahner in this respect. Ramsey cites Rahner in "there is nothing possible for man that he ought not to do," and "evil is the absurdity of willing the impossible."[49] He does not appreciate or explore what Rahner means by possibility or impossibility. Impossibility is that which has no being or meaning, that which is outside a relationship with God. It is directed toward nonbeing rather than the good. In the long term, what is immoral is impossible because as measured against the whole of reality it is impotent. Possibility, on the other hand, is in the theological sense that which is directed toward being, in relation to God. Rahner was attempting to mediate between a strong moralist position that suggests that there are some things that humanity ought never to do and a sceptical position that suggests that one cannot stop humanity doing what is possible. The language he used in his solution to this problem was philosophical and in a literalist sense did not make much sense, but equally one might ask whether Ramsey's theological proclamations about what ought or ought not be done have shown sufficient nuance.

In addition, it is clear that Rahner does have some highly critical comments to make about genetic engineering as such, even though in other areas of his work he resisted blanket condemnation of genetic research.[50] Rahner gives some ethical weight to what he terms the "moral faith instinct" in questions about genetic manipulation.[51] A moral faith instinct means "a universal knowledge of right and wrong belief."[52] He argues that it is justified since the complex nature of the subject is such that it cannot simply be subject to analytical reflection, while at the same time it often cites particular reasons behind a particular judgment. This seems to be a contemporary version of the need in classical thought for the theological virtues of faith, hope, and charity as a prerequisite for the development of other virtues.[53] Rahner is also aware of the dangers of such an approach used in isolation from other forms of reasoning. He links this with the cardinal virtue of prudence in suggesting that there is an element of faith instinct in a "fundamentally synthetic knowledge, formed by the unity of a prudential judgement and a unique moral existential situation."[54] This synthesizing element seems to be of the same character as prudential judgments discussed in more detail earlier. As such, Rahner suggests that the faith instinct has a "right and obligation to reject genetic manipulation."[55] Such a rejection might seem to be stark in light of his more positive assessments about human manipulation cited earlier. Does this mean that Rahner argues in the last analysis against all IGM? In the first place he suggests that all human beings, whether or not born naturally, have to accept the givenness of the world in which they are placed and the giftedness of life in all its forms. Hence the other person must always be viewed as one made and accepted rather than as one chosen or designed. He suggests that genetic manipulation "is the embodiment of fear of oneself, the fear of accepting one's self as the unknown quantity it is."[56] It is clear that once humanity has the illusion of total planning and control, it has ceased to view IGM in the right way.

One might ask, however, given the Human Genome Project and the way each gene codes for multiple functions, could any genetic manipulation inevitably entail this attitude? Also, what if the faith instinct changes, so that it now seems more morally right than it earlier might have that some forms of genetic engineering are acceptable whereas others are not? This seems to be the case with public attitudes to in vitro fertilization. I suggest that Rahner was writing at a time (1968) when genetic engineering was too poorly understood to arrive at adequately sophisticated reasoning, while his notions of transcendental and categorical freedom and his overall positive appraisal of the possibility of manipulation of humans as morally considerable still stand.

Rahner does, nonetheless, raise some important issues that are highly relevant in relating the ethics of IGM to the virtues. In this it is a more helpful alternative than the somewhat stale debate about drawing a moral line between enhancement and therapy, which is morally ambiguous anyway.[57] For example, Rahner is conscious of the need to explore the motive behind genetic manipulation, to analyze its justification in depth, and where appropriate, to challenge the rationale offered if it seems to be disingenuous. I would argue that he is, however, incorrect to assert, which he seems to do, that genetic manipulation is driven by fear of one's destiny. In arriving at this conclusion he believes that in accepting genetic engineering, humanity is accepting what cannot be predetermined. By this he seems to mean humanity is rejecting the ultimate predetermined nature of human existence as given by God. However, this assumes that planning and design are inevitably integral to all genetic engineering, which I suggest they are not.[58] Would all such genetic changes lead to this attitude, to this desire for design to the extent that humanity is no longer considered a free gift? I suggest that this is part of the problem of including all genetic engineering under one umbrella. If it were to be used, for example, to prevent a lethal disease, this would amount to calling into life a person who would not otherwise exist. Is a predetermined destiny toward nonexistence morally good? If we follow this line, then medical interventions at the start of life are illicit, rather than being a gift from God. Given the number of premature births of healthy children currently possible, such an attitude toward medical intervention can no longer be justified.

This is not to deny the real and unique moral significance of crossing the boundary so that the human germline is changed. I suggest that a great deal of care needs to be paid to the extent to which removing a disease trait might imply a judgment about those who are suffering from various diseases and disabilities. Some diseases, in other words, might be morally more ambiguous than others. While moral theologians have often placed an ethical boundary between removal of deleterious genes and enhancement, I suggest that this is too loose a boundary at present; for example, who is to say that very weak intelligence identified as genetic in origin is not in some sense a genetic disease?[59] Those who suffer from Down syndrome, for example, clearly cannot be held to have a life not worth living, yet it is one of the conditions that is routinely screened out through prenatal diagnosis, where relatively late terminations amount to the only form of treatment.[60] The pressure to include such conditions on the list of conditions liable to genetic manipulation would be great and given that this condition arises out of the presence of an extra chromosome, it might be relatively easy to change, for example, by targeting and deactivating one of those chromosomes.[61] Medical practitioners will of course reject the idea that attempting to remove disease traits from the human genome is a judgment about those who suffer from such diseases.[62]

Rahner also objected to artificial insemination using donor sperm (AID) on the basis that this tore asunder the act of love in procreation and its inner relationship to the child. In this, however, he was rejecting not simply AID, but all forms of artificial conception and procreation. Clearly, genetic engineering will inevitably involve a beginning that includes IVF, with or without human cloning. Leaving aside the latter possibility for the moment, I suggest that genetic manipulation should not be viewed as necessarily harmful to human relationships and intimacy in the way he seems to suggest.[63] However, the likely success of any of the technologies does need to be taken into account. The women who come forward for IVF treatment are often vulnerable and liable not to hear or fully understand the statistical nature of the risks involved, the emotional upheaval IVF brings, or the chances of success, which progressively decrease with increase in maternal age. Any approval of IGM needs to take into full account the real needs of the women who are going to be involved. They are not simply disembodied wombs into which newly fertilized eggs are to be implanted, but people with real needs and aspirations. There is a tendency to answer such an ethical issue by bland remarks about the need for adequate counseling.

Rahner believed that even given his understanding of the relationship between categorical freedom and transcendental freedom, the human requirement for genetic manipulation is not compelling enough to justify its use. For this argument he uses examples of increased intelligence or extending life expectancy, rather than removal of genetic disease. It is one reason, perhaps, why Rahner's analysis is less useful than it could have been because he seems to have been ignorant of the scientific possibilities. He suggests, for example, that it is "vital for humanity to develop a resistance to novel possibilities."[64] In addition he was understandably worried about state (eugenic) control over genetic engineering. However, in a democratic society one might anticipate that laws that allow some forms of genetic manipulation to take place and not others would be beneficial for society as a whole. A total withdrawal of any state involvement in the regulation of genetic manipulation would leave genetic practice to the whim of market forces and scientific curiosity. Thus while I agree that state control of genetic engineering is undesirable, its regulation is fully warranted and necessary for the overall health of a community. Rahner also acknowledges that not all the human race could be manipulated by genetic engineering, but for him this does not thereby justify even its limited use.[65] His suggestion that there might be a super race of genetically engineered humans living alongside those who have not been so manipulated is the stuff of science fiction, epitomized in contemporary films such as *GATTACA*. Other science writers have also echoed his fears.[66] As I have argued here, ethically responsible IGM must include the option to say no to some developments.

Rahner mentions the virtues of renunciation and sacrifice as lessons to be learned in relation to genetic manipulation. I would agree that there are some areas that Christian theologians would want to designate as being unacceptable from a theological perspective. This may entail a form of renunciation, an acceptance of suffering as being part of the human condition. However, in addition to these negative aspects, I suggest that the virtue of temperance can include a right understanding about oneself that allows some areas to be justified and not others. Temperance includes humility, a humility that is wondrous in the face of the discoveries of genetic science, but is suitably hopeful about its possibilities. Rahner cannot move to a position where some applications of genetic manipulation rather than others are acceptable, as he has labeled such change in practice under a negative category of design, even while adopting a more open rhetoric based on freedom. While I can agree that some technologies might in the long term have dehumanizing tendencies, it seems to me that eliminating the possibility of genetic change being used for the good of humanity and under the guidance of the Holy Spirit is to limit God-given possibilities for the future. I would not want to go as far as suggesting that humanity becomes co-creators through such technologies. Language such as this might encourage a false sense of hubris, and it is not worth taking this risk. Rather, in examining the overall trends offered through the technology, alongside particular cases, judgments need to be made as to which decisions are prudential and which are not.[67]

Given that I have argued that in principle IGM cannot be ruled out, it is necessary to consider not just the goals that are implicit in such a change but also the means and the likely outcomes. The means, for example, might include human cloning using cell nuclear replacement.

In this technology, an embryo that is known to carry a deleterious gene would be taken to the stem cell stage. The stem cells would then be genetically manipulated and the resulting tissue would be used as the nuclear source for a cloned embryo in which another enucleated egg from the mother was used. This would lead to a healthy cloned human being through the creation of an early disease-bearing embryo. Ian Wilmut has endorsed this technique as an acceptable use of human reproductive cloning.[68]

I suggest that this method for IGM is wholly unacceptable for a number of reasons. In the first place, its development opens the way for human reproductive cloning, which is highly controversial from a theological point of view.[69] Second, the human embryo clearly becomes a means to achieving an end, namely the cloning of another human being free of disease, genetically identical in all respects apart from the deleterious gene. Although there may be little or no life expectancy for such an embryo, to treat it as a source of human life where life does not exist is morally distinct from using embryonic stem cells, for example, to save the life of someone who is dying of that disease. Even the use of therapeutic cloning to treat disease in adults or children is morally ambiguous. The extent of its moral ambiguity will reflect the status given to the early embryo. While I suggest that as an interim measure some use of unviable spare IVF embryos to generate stem cells for treatment of disease may be justified, to use an embryo, even a spare IVF embryo, to generate a new individual through human cloning crosses another moral boundary.

Other alternatives have been suggested that create sperm or eggs using embryonic stem cells and this is now reported in mice.[70] In these cases the sperm or egg could be manipulated. However, the means is still faulty; that is, the source of embryonic stem cells has entailed deliberate creation and then destruction of an embryo for this purpose. What if this is the only means of achieving IGM? I suggest that even if this is the only means available, it needs to be resisted until such time as scientists have discovered alternative methods, perhaps through manipulation of the egg or sperm cells prior to fertilization. There may be other ways of generating eggs and sperm from stem cells inherent in the gonads themselves, and these may be a possible source for IGM in the future.[71] The method of modification that involves gene repair rather than gene replacement is also one that would seem to be more acceptable because it would entail less drastic change and thus, would not upset the delicate regulatory apparatus existing in cells.

## Conclusions

I began this chapter by exploring the notion of freedom and knowledge, asking whether there might be areas of knowledge that are unacceptable to pursue from a theological point of view. An exploration of the story of the fall of humanity in which the knowledge of good and evil is named as the outcome of eating the forbidden fruit implies less that certain forms of knowledge are impermissible and more that a good life is one that is lived in covenant relationship with God, rather than outside that covenant. The mythological account is nonetheless significant in its reminder that humans are inevitably fallible in how they justify their actions and believe that something is good when their desires are misdirected toward their own ends. From this perspective, I explored different theological interpretations of freedom, stemming from either a stress on freedom of choice among unlimited possibilities or freedom that arises through gaining a particular skill or living in relationship with God. The latter, freedom for excellence, integrates various aspects of freedom that freedom of indifference separates and also provides a way of encouraging a positive approach to scientific developments while being aware of nascent dangers and difficulties. I also compared this classical view with Rahner's understanding of the fundamental option, to live a life in orientation toward God, a life lived out of transcendental freedom expressed as a decision of love and leading to the categorical freedom in which specific decisions can be made.

Of course, once we consider specific decision making in the context of freedom, notions of conscience spring to mind. What does it mean to act out of one's conscience? Augustine was more conservative than Aquinas in this respect and argued that it was possible for humanity to be mistaken in it conscience, so its leanings needed to be rejected. Aquinas, on the other hand, argued that as long as persons were convinced that their action was good, then it made no sense to interpret their action of conscience as against God's laws. He was of course more optimistic about humanity's capacity for good compared with Augustine, who tended to put more emphasis on the doctrine of original sin. However, I suggested that Aquinas' understanding of conscience was best situated within his more developed sense of prudence, so that an act of conscience was the judgment made about those actions arising out of prudential decision making. Prudence, as one of the four cardinal virtues, is relevant in that it includes the capacity to bring together various facets of human understanding, including, for example, memory, circumspection, foresight, caution, and taking counsel. It is through careful deliberation, judgment, and action that particular decisions are made in an ethically responsible way. Prudence, like wisdom, may be learned, but it can also be a gift from God in much the same way that we find dialectic between freedom of choice and transcendental freedom in Rahner.

When it comes to more specific questions about which forms of IGM to accept or reject, I challenged the arguments against genetic engineering that Ramsey used, namely playing God, and those that Rahner used, most particularly that it represented humanity as a designer, rather than accepting its place as predetermined by God. There are, nonetheless, aspects of Rahner's argument that are worth careful consideration. His desire to look at the motives of those involved is important, as well as his more general affirmation that in principle the possibility of manipulation of humans needs to be considered. His work does appear contradictory in certain respects. It may be that the knowledge of the science available at that time was still rudimentary, suggesting the importance of a fully informed discussion by moral theologians.

While the details of what might be feasible in the future have yet to unfold, I have argued for the admissibility of very limited use of IGM for lethal genetic conditions, ideally using methods of repair, where the risk of side effects is thought to be slight and where other possibilities have been discounted. In addition I have argued for a relatively tight boundary at least for now, not just excluding interventions that fall within the somewhat fluid category of enhancement but ruling out the use of these technologies for anything short of lethal diseases. I have also argued against the use of human cloning or embryonic stem cell technology in order to achieve the aims of IGM, where this would entail human reproduction through a cloning method. Nonetheless, since the technology is rapidly developing all the time, the ethical debates are never likely to stand still—there is a continual need for both wisdom and prudence in ethical decision making, where wisdom represents a shared search, one that is not closed to the possibility of change or afraid to challenge new developments where appropriate.

## Notes

1. I prefer to use the term "inherited genetic" rather than "germline" because the former is more inclusive of a wider range of techniques available.

2. This term was popularized by Francis Fukuyama in *Our Posthuman Future: Consequences of the Biotechnology Revolution* (London: Profile Books, 2002).

3. For a discussion of this, see Anne McClaren, *Cloning (Ethical Eye)* (Strasbourg: Council of Europe, 2002).

4. For an excellent overview of the various factors affecting views about the status of the human embryo and their relationship to ethical issues in genetics, see Brent Waters and Ronald Cole-Turner, eds., *God and the Embryo: Religious Voices on Stem Cells and Cloning* (Washington, DC: Georgetown University Press, 2003).

5. Cited in Claus Westermann, *Genesis 1–11*, tr. J. Scullion (Lanham, MD: Rowman & Littlefield, 1995), p. 222.

6. Westermann, *Genesis 1–11*, p. 223.

7. Westermann, *Genesis 1–11*, p. 223.

8. In this I am following Westermann's commentary, *Genesis 1–11*, p. 277.

9. From an ecological perspective we might even see this as the start of distinctive human behavior, a sense of difference from the animals and other creatures, which would know nothing of the emotion of shame in being naked.

10. K. Rahner, *Theological Investigations*. vol. 19, *Faith and Ministry*, trans. E Quinn (London: Darton Longman and Todd, 1983), p. 17.

11. S. Pinckaers, *The Sources of Christian Ethics*, trans. M. Noble (Edinburgh: T. & T. Clark, 1995), p. 338.

12. This list is a modified version of that found in Pinckaers, *Sources of Christian Ethics* p. 350.

13. I am borrowing this term from Pinckaers, *Sources of Christian Ethics*, pp. 253 ff.

14. Aquinas, *Summa Theologiae*, vol. 11, *Man*, trans. T. Suttor (London: Blackfriars, 1970) 1a, Qu. 75–76, 79, 80–83; and *Summa Theologiae*. vol. 34, *Charity*, trans. R. J. Batton (London: Blackfriars, 1975) 2a2ae, Qu. 24.

15. Karl Rahner, *Theological Investigations*. vol. 11, *Man in the Church*, trans. Karl-H. Kruger (London: Darton Longman and Todd, 1963), pp. 235–264.

16. Karl Rahner, *Theological Investigations*. vol. 6, *Concerning Vatican Council II* (London: Darton Longman and Todd, 1982).

17. Karl Rahner, *Sacramentum Mundi*. vol. 2, *Theological Freedom* (London: Burns and Oates, 1968) p. 361.

18. Rahner, *Sacramentum Mundi*. vol. 2, *Theological Freedom*, p. 362.

19. Rahner also argues that mystical experience is possible outside the Christian community. K. Rahner, *Theological Investigations*. vol. 18, *God and Revelation*, trans. E. Quinn (London: Darton Longman and Todd, 1983), p. 182.

20. K. Rahner, *Theological Investigations*. vol. 20, *Concern for the Church*, trans. E. Quinn (London: Darton Longman and Todd, 1981), p. 54.

21. Rahner, *Theological Investigations*, vol. 20, *Concern for the Church*, p. 54.

22. This idea of radical openness and incompleteness is one used by Rahner in relation to humanity's task of self-determination, given by the radical gift of freedom, "laid upon him as a burden." See K. Rahner, "The Experiment with Man," in *Theological Investigations*. vol. 9, *Writings of 1965–7, part 1*, trans. Graham Harrison (London: Darton Longman and Todd, 1972), pp. 212–213 (full reference pp. 205–224).

23. See also the discussion in K. Rahner, "The Problem of Genetic Manipulation," in *Theological Investigations*. vol. 9, *Writings*, pp. 225–252, especially p. 228.

24. Ted Peters, *Playing God: Genetic Determinism and Human Freedom* (London: Routledge, 1997).

25. For a critical discussion of trends in bioethics, see Julie Clague, "Beyond Beneficence, The Emergence of Genomorality and the Common Good," in C. Deane-Drummond, ed., *Brave New World? Theology, Ethics and the Human Genome* (London: Continuum, 2003), pp. 189–224.

26. Aquinas did not use this term, it is Rahner's, but it is a helpful distinction among the facets of freedom, although one should not be artificially separated from the other.

27. Bernard Häring developed the idea of the fundamental option being the basic orientation of a person's life. See, for example, B. Häring, *Free and Faithful in Christ*, vol. 1 (New York: Seabury, 1978), *The Law of Christ*, vol. 1 (Cork, Ireland: Mercier Press, 1960). For further discussion of the concept of freedom in relation to conscience, see Linda Hogan, *Confronting the Truth: Conscience in the Catholic Tradition* (London: Darton Longman and Todd, 2001), pp. 128–135.

28. Aquinas, *Summa Theologiae*. vol. 28, *Law and Political Theory*, trans. T. Gilby (London: Blackfriars, 1966), ia2ae, Qu. 94.1.

29. Hogan, *Confronting the Truth*, pp. 18–20.

30. An example is found in 1 Timothy 1.5. These phrases suggest synonymous meanings; that is, conscience is associated with having moral integrity. I am grateful to Rudolf Heim for this observation.

31. Brian Davies, *The Thought of Thomas Aquinas* (Oxford: Clarendon Press, 1992), p. 235.

32. Hogan, *Confronting the Truth*, pp. 51–62.

33. Aquinas, *Summa Theologiae*. vol. 38, *Injustice*, trans. Marcus Lefebure (London: Blackfriars, 1974), 2a2ae, Qu 79.13.

34. Aquinas, *Summa Theologiae*. vol. 18, *Principles of Morality*, trans. Thomas Gilby (London: Blackfriars, 1965), Ia2ae, Qu. 19.5.

35. Daniel Westberg, *Right Practical Reason: Aristotle, Action and Prudence in Aquinas* (Oxford: Clarendon Press, 1994), p. 105.

36. For further discussion on prudence, see C. Deane-Drummond, *The Ethics of Nature* (Oxford: Blackwell, 2004), pp. 10–15.

37. The traditional formulation takes its bearings from the virtues possessed by the coming Messiah, as described by Isaiah11.1–2, and taken up in the earliest Christian traditions as corresponding to the gifts of the Spirit. See Romanus Cessario, *Introduction to Moral Theology* (Washington, DC: Catholic University of America Press, 2001), pp. 205–212.

38. In light of this, it might seem surprising that some moral philosophers have tended to abstract the idea of natural law from Aquinas and ignored its link with covenant relationships. For discussion see Deane-Drummond, "Wisdom Ethics, Aquinas and the New Genetics," in C. Deane-Drummond and B. Serszynski, eds., *Re-Ordering Nature: Theology, Society and the New Genetics* (London: T. & T. Clark/Continuum, 2003), pp. 293–311.

39. I am grateful to Rudolf Heim for a discussion on the place of conscience in the context of prudential decision making in the thought of Thomas Aquinas.

40. Westberg, *Right Practical Reason*, pp. 7–8.

41. Peter Vardy, *Being Human: Fulfilling Genetic and Spiritual Potential* (London: Darton Longman and Todd, 2003), p. 71.

42. Of course, if a cavalier approach to abortion is taken, then such "mistakes" could be screened during pregnancy, but this has its own ethical difficulties.

43. Aquinas, *Summa Theologiae*. vol. 36, *Prudence*, trans. T. Gilby (London: Blackfriars, 1974), Qu. 49. For further discussion on prudence, see C. Deane-Drummond, *The Ethics of Nature* (Oxford: Blackwell, 2004).

44. Paul Ramsey, *Fabricated Man: The Ethics of Genetic Control* (New Haven, CT: Yale University Press, 1970).

45. I have argued elsewhere that the most promising method in this case is gene repair because it involves no new genetic material being introduced. For discussion see C. Deane-Drummond, *Genetics and Christian Ethics* (New Studies in Christian Ethics) (Cambridge: Cambridge University Press, 2006).

46. Ramsey, *Fabricated Man*, p. 45.

47. Ramsey has in this sense been characterized as being more negative in his attitude to the possibility of genetic engineering than is actually the case. See discussion in Peters, *Playing God*, pp. 143–144.

48. Ramsey cites Rahner as suggesting that everything is always legitimate; see Ramsey, *Fabricated Man*, pp. 139–140.

49. Ramsey, *Fabricated Man*, pp. 140–141.

50. The ambiguity in Rahner's thought, that is, the contrast between his positive statements on genetic manipulation and the need to resist fear of change, and his other more critical remarks on genetic manipulation were not apparently noticed by Paul Ramsey, who assumed he endorsed most aspects of genetic engineering, a view that also seems implicit in Cole-Turner. [R. Cole-Turner, "Human Limits: Theological Perspectives on Germ Line Modification," in Audrey R. Chapman and Mark S. Frankel, eds., *Designing Our Descendants: The Promises and Perils of Genetic Modifications* (Baltimore: John Hopkins University Press, 2003), p. 196.] It is also implicit in the way that Ted Peters refers to Rahner in *Playing God*, pp. 143 and 156.

51. Rahner's critical discussion was directed toward the particular practice of AID, while he left open the possibility of experimentation with embryos.

52. Rahner, "The Problem of Genetic Manipulation," pp. 238–243.

53. See Cessario, *Introduction to Moral Theology*.

54. Rahner, "The Problem of Genetic Manipulation," p. 240.

55. Rahner, "The Problem of Genetic Manipulation," p. 243.

56. Rahner, "The Problem of Genetic Manipulation," p. 245.

57. President's Council on Bioethics, *Beyond Therapy: Biotechnology and the Pursuit of Happiness* (New York: Dana Press, 2003), pp. 15–19.

58. This debate is also discussed helpfully by Ronald Cole-Turner, "Human Limits," in Chapman and Frankel, eds., *Designing Our Descendants*, pp. 188–198.

59. Neil Messer discusses some of the issues surrounding the difference between enhancement and removal of deleterious genes in "The Human Genome Project, Health and the Tyranny of Normality," in Deane-Drummond, ed., *Brave New World?*, pp. 91–115.

60. It might be argued that preimplantation genetic diagnosis is more acceptable in this respect, but then this depends on a view about the moral status of the embryo and the question of whether this might negatively affect those suffering from the disease by generating a social context that is less supportive of those who decide to go ahead with the pregnancy.

61. There has been some discussion of the treatment of those with Down syndrome through somatic gene therapy. However, early hopes that the disease might be treatable have been challenged by more recent experiments on mice,

which suggest that Down syndrome is not limited to short sections of the chromosome, as was once thought to be the case.

62. J. Bryant and J. Searle, *Life in Our Hands: Christian Perspectives on Genetics and Cloning* (Leicester: IVP, 2004).

63. Rahner's views are somewhat ambiguous here because he uses the language of genetic manipulation in referring to heterologous artificial insemination and although he speaks in places about "test-tube" babies, he was writing before IVF became readily available; see Rahner, "The Problem of Genetic Manipulation," p. 244.

64. Rahner, "The Problem of Genetic Manipulation," p. 249.

65. Rahner, "The Problem of Genetic Manipulation," p. 247.

66. See, for example, Bill McGibben, *Enough: Genetic Engineering and the End of Human Nature* (London: Bloomsbury, 2003).

67. I am using prudence in the theological sense, that is, according to Thomistic ethics.

68. Ian Wilmut cited in S. P. Westphal and P. Cohen, "Cloned cells today: Where tomorrow?", *New Scientist* 181 (February 21, 2004): 6; Ian Wilmut, "The moral imperative for human cloning," *New Scientist* 181 (February 21, 2004): 16.

69. A discussion of all the ethical arguments against reproductive cloning is beyond the scope of this chapter. For a summary, see Deane-Drummond, *The Ethics of Nature*, pp. 117–129.

70. Sperm or egg stem cells can be derived from the gonads themselves, or more controversially, by using cell nuclear transfer techniques. See R. Nowak, "Stem cells allow infertile to become fathers," *New Scientist* 180 (December 13, 2003), reported in *The Times*, December 11, 2003, "Artificial Sperm Signal End to Male Fertility," p. 14.

71. This has been suggested by Kenneth Culver. See K. W. Culver, "Gene-Repair, Genomics and Human Germ Line Modification," in Chapman and Frankel, eds., *Designing Our Descendants*, pp. 77–92.

# 9

# Religion, Genetics, and the Future

Ronald Cole-Turner

So strong is the religious commitment to healing that germline modification for therapeutic purposes cannot be ruled out by many religious institutions, leaders, and scholars. The contributors to this volume, who represent various Jewish and Christian perspectives, come to a similar conclusion. With one exception, chapters 2 through 8 agree that if strict conditions are observed, the core idea of human germline modification cannot be ruled out on religious grounds. Whether the conditions can be met—indeed, whether they must all be met in order to proceed—is not yet clear in these discussions. What is clear, however, is that most religious voices considered in this volume leave the door open to the moral possibility of modifying the human germline.

This final chapter explores the conditions of acceptability more fully. First, however, a summary of the religious argument in support of human germline modification is offered. In the second section, four moral or religious conditions are considered. These include safety, protection of embryos, a concern for social and economic justice, and support for therapy versus enhancement. In the final section, attention is turned from technology to those of us who, whether we wish to or not, must live in a world where technology reshapes human life. Emerging technologies, such as germline modification but now also including a growing range of other strategies, offer us the power to change humanity according to our desires. If so, then we should examine these desires and the moral frameworks from which they arise. What is it we value most deeply about ourselves, that we would wish to enhance in our offspring? The final section is an invitation to reflect on these themes.

## Religious Support for Human Germline Modification

Perhaps the best way to understand religious support for germline modification is by reflecting first on the moral debate about a technology that is already available today. This technology, known as preimplantation genetic diagnosis (PGD), allows couples using in vitro fertilization to create multiple embryos and then test each embryo for specific genetic conditions. To many observers, PGD seems like a precursor to human germline modification. Using PGD, couples select embryos free of a specific genetic disease. Using germline modification, however, couples might go one step further to create a genetically modified embryo that is free of a specific disease. In both cases, the goal is the same, to start a pregnancy without a higher than normal risk of a specific genetic disease.

Scientifically, PGD is much easier and safer than germline modification because it involves no risky or unpredictable genetic changes. Germline modification, on the other hand, might someday allow couples to engineer modifications, which is a far more powerful strategy than simple selection from among available embryos. Some people today already refer to PGD as creating "designer babies," but the real element of design and all the concerns about enhancement become possible only with the future arrival of germline modification. For that reason, some people who accept PGD are opposed to germline modification because they fear the greater power it offers. Others, including many of the religious voices included in this volume, argue just the opposite: PGD is morally flawed, but germline modification, if safe and limited in scope, is morally preferable and acceptable.

The objection to PGD is that it creates multiple embryos in order to select some for destruction. The justification for the procedure, of course, is that it allows couples at risk for genetic problems to start a pregnancy conceived by them while reducing the risk of passing along a genetic disease. For many, that is justification enough to warrant the procedure. For others, the justification is strong but the moral complications are still worrisome. For example, LeRoy Walters and Julie Gage Palmer write that "prenatal diagnosis followed by selective abortion and preimplantation diagnosis followed by selective discard seem to us to be uncomfortable and probably discriminatory halfway technologies that should eventually be replaced by effective modes of treatment."[1]

A more somber and prohibitive note about PGD is sounded by Jürgen Habermas, who writes that "many of us seem to have the intuition that we should not weigh human life, not even in its earliest stages."[2] Underneath these concerns lie the burdens of history and of medicine without moral constraint: "[T]he fact that we make a highly momentous distinction between life worth living and life not worth living for others remains disconcerting."[3] In the case of PGD, selection and destruction is a "binary decision" that "already betrays an intention to improvement. The selection is based on a judgment of the quality of a human being."[4]

It should be said again, in anticipation of the discussion later in this chapter, that one moral advantage of PGD over germline modification is that PGD is severely limited in its power to offer enhancement capabilities. The limiting factor here is the number of embryos that can be tested in any one case. When PGD technology is applied to a set of eight to twelve embryos, the number of different genetic tests that can be run is limited since each test excludes a significant portion of the embryos. Unless more embryos can be created and tested at one time, it is unlikely that PGD can be used to test for a disease plus test for genes that might enhance other traits, especially complex, multigene traits like cognitive ability. It may be possible in the future to create more embryos for testing, but this compounds the moral problem of creating embryos only to destroy them.

By comparison with PGD, human germline modification does not create multiple embryos and then select the healthy ones for life. It seeks to create a healthy embryo in the first place. It can be entirely therapeutic, at least theoretically, not just in regard to the limits of its aim but also in its consideration of each human embryo. To the extent that this is true, it requires religious support or at least religious acceptance. PGD draws the condemnation, for instance, of the Roman Catholic Church, but human germline modification receives a qualified endorsement. In addition to the statements quoted in chapter 1 and adding further to the comments of Thomas A. Shannon (chapter 3) and James J. Walter (chapter 6), the statement of a distinguished Catholic bioethicist, Albert Moraczewski, should be noted. After his careful review of Catholic teachings on germline modification, Moraczewski concludes "that such intervention on human germ-line cells if done with a clear therapeutic intent could be morally acceptable provided that the process met certain conditions: *if* the process did not destroy or impede essential components and processes of human nature, such as the capacities to know and love humanly, and *if* other issues such as safety, efficacy, and free, informed consent of future generations could be resolved. Another major caveat is that the means employed to insert the gene into the gamete and subsequent fertilization should not involve IVF or other procedures that the Church deems to be contrary to the dignity of the resulting human being and the sacredness of human procreation."[5] These restrictions are important and will be considered more in the next section of this chapter.

A recent book by a conservative Protestant, Edwin C. Hui, comes to the conclusion that germline modification is "a morally risky undertaking that can only be endorsed as the means to prevent or remedy disorders that would otherwise result in great suffering and early death."[6] Hui is concerned especially about how "germ-line gene therapy can so easily move toward eugenic enhancement intervention reminiscent of the Nazi effort that one nation or a small alliance of nations should not be allowed to monopolize and control this technology. Instead a dialogue about genetic intervention that involves future generations should include a global participation of all nations of the world because what is at stake is nothing less than the future of humanity as a whole."[7] While one might agree with Hui's call for global dialogue, it is not clear why a global as opposed to a national decision is more likely to result in a morally acceptable outcome, unless it is to head off some sort of national competition to enhance the cognitive performance of future citizens. As much as Hui fears enhancement applications, he does not argue that the prospect of enhancement is so serious that it forces us to resist the technology.

Sondra Wheeler, a Protestant moral theologian, likewise rejects some hypothetical uses of germline modification but approves of others: "Seeking to select the genetic characteristics of our offspring in accord with cultural values or parental preferences is incompatible with honoring the dignity of a creature whose source and destiny is in God. . . .

Therefore, genetic interventions aimed at increasing or enhancing positive characteristics, even real goods such as intelligence or creativity, cannot be defended as essential to well-being and should be forgone."[8] Nevertheless if it is true that germline modification is the only way to avoid some forms of grave illness, it might be acceptable, Wheeler argues. "If all the concerns for the reliability of correction, insertion, expression, and inheritance of genetic material can be addressed, and the safety of such limited changes in the gene pool assured to a level comparable with the known risks of leaving such defects unaddressed, I see no absolute barrier to such interventions in the limits of human stewardship."[9]

The religious view shared by many Christian and Jewish commentators is that germline modification is morally acceptable within certain limits. The first condition or limit, which is shared by nearly everyone who ponders the question, is that human germline modification should not be attempted until there is a reasonable level of technical certainty that it can be done at an acceptable level of safety. Beyond the safety condition, however, three other relevant conditions are attached to any moral approval of germline modification. The scholars and the religious traditions in which they are situated might differ in their endorsement of these three conditions or even on how to define them. The three additional conditions are avoidance of harm to embryos, avoidance of applications that are likely to increase injustice, and avoidance of enhancement. These four conditions or qualifications, taken together, are explored more fully in the next section.

## Four Conditions That Limit Religious Approval

Religious support for human germline modification is not unanimous, but the majority of authors in this volume and the majority of the opinions of other authors and texts, to reiterate, leave the door open to the possibility of this technology. Rarely do they say that developing this technology is a moral priority. More important, however, are the moral conditions attached to approval. Unless one or more of these conditions is met, approval is withheld. This section reviews those conditions and explores the possibility of their being met.

**Avoid Unacceptable Levels of Risk**

That human germline modification must be shown to be safe before it is used is a demanding but obvious moral condition that is shared by nearly every commentator on the subject, religious or not. Religious scholars introduce no special standard here, nor do they require that technology must guarantee a perfect standard of safety before it is morally permissible to proceed. Furthermore, religious scholars recognize that the level of safety is largely a technical question to be assessed by experts in the relevant scientific and medical fields and in such areas as medical research and the ethics of clinical trials.

The question of the effects of germline modification on future generations raises a special safety concern. It may turn out that safety problems are only recognized in the distant future, presenting problems that are not foreseen today. While religious scholars recognize this possibility, and while none of them intentionally minimize its seriousness, they rarely focus on it. Almost no one raises it as a special objection to germline modification. A notable exception is a 1992 statement by the United Methodist Church: "Because its long-term effects are uncertain, we oppose genetic therapy that results in changes that can be passed to offspring (germ-line therapy)."[10] If anything, however, religious institutions and scholars are less concerned than their secular counterparts about the problem of long-term safety.

**Avoid Harm to Human Embryos**

Official Roman Catholic moral guidelines prohibit any form of reproductive technology that creates embryos outside the human body, such as in vitro fertilization. Furthermore, any instrumental use of the human embryo is prohibited. One embryo cannot be used or destroyed to benefit another, even to benefit many others. These guidelines are reviewed in chapter 3 by Thomas Shannon.

In light of these long-standing principles, official Catholic approval of human germline modification is contingent upon avoiding harm to human embryos. This means that germline modification is unacceptable if it uses some embryos as a resource or a method to benefit others. In addition, the DNA of the embryo cannot be modified while the embryo is outside the body. In effect, this means that official Catholic approval is dependent upon future technical advances that will permit the modification of DNA, not in the embryo once it is created, but in the human sex cells, such as eggs or sperm, prior to the creation of the embryo. Furthermore, these modified sperm or egg cells will have to be joined by sexual intercourse, not by reproductive technology. Recent advances suggest that while these conditions are technically demanding, they may not be impossible. If they are met, then official Catholic approval of human germline modification would appear to be forthcoming, based on the statements that are reviewed in chapter 1.

This moral condition is grounded in a long-standing Christian view of the value of early human life. This view, which is not shared equally by all Christians, is historically rooted in Hebrew thought. From its beginnings, Christianity has been far more protective of the embryo than Judaism, largely because of the Christian doctrine of the incarnation, which holds that God is personally present in Jesus Christ from the beginning of his conception. For Catholics more often than other Christians, the conception of Jesus within Mary is especially important in popular piety and therefore in theology, giving rise to the view that Christ's embryo from the beginning is seen as a vessel of the divine.[11] One interpretation of this doctrine means that all human embryos must be treated as the moral equivalent of full human lives. Therefore, not only must germline modification be therapeutic in its goal, it must also be therapeutic in its means to that end, treating every embryo involved in the process as a full human subject, not as an experiment or a clinical resource. Not all Christians, indeed not all devout Roman Catholics, accept this view of the embryo or hold to this condition.

If accepted, however, this moral condition place limits on the technology of human germline modification in three critical ways. First, the methods used to achieve modification must avoid any instrumental use of embryos, such obtaining stem cells from human embryos. Any effort to produce eggs or sperm from embryonic stem cells (see chapter 1) is precluded, at least as long as the cells are literally derived from an embryo that is destroyed in the process. If it becomes possible to derive pluripotent stem cells, the functional equivalent of embryonic stem cells, from sources other than human embryos, this limit might be set aside.

Another possibility is that eggs or sperm might be created using stem cells from adult sources.

Second, even if the genetically modified eggs or sperm might be created in a morally acceptable way, in vitro fertilization could not be used to create the new, genetically modified embryo. It might be possible to meet this demand by genetically modifying the cells that produce sperm. Once these sperm-producing cells are inserted into a male body, they might generate genetically modified sperm that act as normal sperm except that when they fertilize an egg the result is a genetically modified embryo.

Third, any use of preimplantation genetic diagnosis or prenatal genetic testing with a view to the possibility of terminating a pregnancy, which will surely be required in order to test any attempts at germline modification for success, is also ruled out. Without these standard methods of testing an embryo or a fetus, it is unlikely that physicians and laboratories will agree to offer germline technology in the first place. In the end, this limit may be the most difficult of all. What is desired from the point of view of reproductive technology, and what is flatly prohibited by official Catholic teaching, is a quality-control test that will offer reasonable assurance that the genetic modification will be effective and that it will not create complications. Without the ability to screen for technologically introduced errors, no reproductive specialist will want to offer germline modification.

Shannon reviews the official Catholic teachings on the human embryo and embryonic stem cell research. His own position offers a different view of the early embryo based upon well-grounded traditional Christian and Catholic perspectives, but it is at odds with the current teaching of the Magisterium. Shannon's view allows such things as human embryonic stem cell research and removes the three restrictions noted in the previous paragraph. Although they are not the official teachings of the church today, Shannon's perspectives enjoy a solid base in traditional theology and may be shared by many individual Catholics, not to mention others who share with him a concern to protect nascent life but disagree about the centrality of conception as its moral marker.

Among Protestants, there is no general agreement on the question of the moral status of the embryo and about such things as embryonic stem cell research, in vitro fertilization, preimplantation genetic diagnosis, or amniocentesis. As a rule, these things are permitted or left to individual conscience, although in recent decades there has been a noticeable trend toward greater caution and restriction. Conservative or traditional Protestants who agree with Catholic teaching about the value of the embryo may in time come to agree explicitly with the official Catholic condition placed on approval of germline modification.

**Avoid Increasing Injustice**

Several contributors to this volume point out what might be called the justice objection or justice condition that should be addressed as part of any approval given to germline modification. It is often observed that nearly every new technology is expensive and its development (at least at first) is beneficial to the wealthy or powerful, not the poor or the weak. While many agree with this observation, they do not conclude that all new technology is morally suspect because it is not distributed evenly or universally. Human germline modification, however, might be present a unique problem in that it turns today's economic advantages into tomorrow's genetic advantages, extending today's gaps into new, more difficult, and more permanent dimensions.

In chapter 7, Lisa Sowle Cahill discusses the theme of justice at length, drawing on the rich backdrop of traditional Catholic teachings on the common good. Cahill calls for the need to denounce and resist any efforts, present or future, to market genetic enhancements as a pathway to social advantage. The prospect of germline enhancement is particularly troubling because it has the potential to increase the very structures of injustice. Not only will the high-end benefits of germline modification (assuming they are realized) be available only to the rich, at least at first, but from that very fact new injustices will arise. The privileged will be able to buy, not just advantages for their children, but advantaged children; not just gifts but giftedness. Lee Silver has speculated that in the distant future two or more human species will result as the genetically enriched breed among themselves and enhance their breeding with technology, leaving the rest of us behind.[12]

A generation earlier, the Catholic theologian Karl Rahner warned "if a partial genetic manipulation became normal practice, consciously recognized by society, it would create two new 'races' in mankind: the technologically manipulated, super-bred test-tube men who inevitably would have a special status in society, and the 'ordinary', unselected, mass-produced humans, procreated in the old way. But what new social tensions would arise from this?"[13] A more recent theological objection based on justice is offered by Audrey Chapman, who concludes that germline modification "would have profound negative societal consequences . . . and would very likely make current injustices and inequalities worse and far more difficult to rectify. . . . From a justice perspective, there seems to be only one option: not to go forward."[14] In other words, germline enhancements will inevitably increase rather than decrease the amount of injustice in the world. For the sake of justice, it must be stopped.

The justice condition draws its primary force from the likelihood that germline modification will be used not just for therapy but for enhancement. Anyone who thinks that enhancement can be avoided is likely to be relatively less concerned about increasing injustice, seeing it as a serious problem but not fundamentally different from current concern about fair access to health care, whether in one nation or on a global scale. On the other hand, those who think enhancement applications will rule the day are most likely to fear that germline modification will make the world less just. Germline enhancement, as some argue, is different from all other technology. The problem is not merely that it will be distributed unevenly. The central problem is that germline enhancement creates new forms of inequity by transforming the current gap in wealth until it creates new divisions based on mental capacity or general healthfulness or longevity, perhaps in combination and perhaps in ways that might never be overcome, sowing new and deeply troubling divisions within the human race.

Can an effective firewall be established between germline therapy and enhancement? If so, then the primary force of the justice argument is removed. In other words, a great deal is hanging on the question of whether we can distinguish therapy and enhancement in a way that has practical force. Cameron and DeBaets argue in chapter 5 that once germline technology is developed, enhancement uses cannot be prevented. If they are right, then the justice condition derives its force from the prediction that germline enhancement will inevitably lead to an unacceptable fracturing of the human community. And in that case, the justice condition leads to the conclusion that the only way to preserve the social fabric of human unity, which is already under threat, is to prevent germline technology altogether.

Then it must be asked: If it is not possible to draw a line around therapy in order to prevent the use of this technology for enhancement, how can it be possible to prevent the development of the technology in the first place? Both actions, and our assessment of whether they are possible, depend upon human global systems of science policy and regulation, which are almost nonexistent. If we lack the power to prevent a technological application, there is no reason to think we possess the power to prevent a technological development.

Even though religious scholars do not agree that the justice concern requires a ban on the development of germline technology, they are concerned nonetheless about what appears to be growing social and economic inequalities in the world and about the role of technology in general in fueling this growth. As a result, their approval of germline modification technology is offered reluctantly. Even if it could be argued that germline modification will not create greater injustice in the world, it is hard to see how its development is a moral priority compared with more urgently pressing health needs around the world, where tens of thousands die each day for lack of basic nutrition or preventive medicine. Some religious writers who offer approval for germline modification do so with a certain wistful reluctance, wishing that exotic medicine would take its place behind the higher priorities of global economic justice.

**Avoid Enhancement**

The final condition, which is the most widely shared among religious scholars, is that germline modification must be limited to therapy and not condoned for enhancement purposes. Few who make this assertion are so naïve as to think that the line between therapy and enhancement is easily drawn, or that once drawn it will be honored. Even so, they argue that most people do in fact have an intuitive sense of a distinction between a disease and a social preference and therefore between therapy that aims to treat or prevent a disease and enhancement that modifies human life according to social preference. A line that cannot be held may still be worth drawing, if only to guide those who seek guidance.

The religious distinction between therapy and enhancement is different from that of its secular counterpart. Some religious thinkers from the western, theistic traditions who employ a therapy–enhancement distinction build their argument on a criticism of the secular version. One problem they note in the secular therapy–enhancement distinction is that it lacks a coherent answer to the libertarian objection. Secular bioethics is centrally based on the core principle of autonomy. Under the banner of autonomy, it is not clear why individual consumers of medical services should not be free to choose the technologies for enhancement purposes. If germline modification is permitted, why stop at therapy?

Comparisons with cosmetic surgery and cosmetic pharmacology come to mind. Medications and procedures, once legal or accepted, may be used for "off-label" purposes, as long as informed consent (autonomy) is respected and other minimal safeguards are in place. According to Brent Waters, "No compelling objections to technological self-enhancement can be offered on late liberal terms so long as three conditions are met: (1) that competent persons have been fully informed and have freely consented to the enhancement methods employed; (2) that no other persons are intentionally harmed in pursuing these enhancements; and (3) that all persons have a fair opportunity to pursue self-enhancement."[15] If it is once agreed that parents can freely decide to use germline modification to create children without a specific disease, indeed that all reproductive decisions are entirely within the scope of parental or "procreative liberty," how will a secular society tell prospective parents they may not use the same technology to create a child with traits they desire?

Religious scholars sometimes note a second problem with secular versions of the distinction between therapy and enhancement. Inevitably, the secular version must search for the therapy–enhancement boundary somewhere in nature, typically in human nature, which is itself under conceptual attack by the very sciences that advance these technologies in the first place. In contrast to the secular form of the distinction, religious scholars tend to agree that the ground for any assessment of enhancement must lie in a theological understanding of the meaning and purpose of human life. The meaning of enhancement might be illumined by a scientific understanding of what is typical or normal for the species, but it cannot in the end be grounded in a study of nature, especially one that claims to be value-free and that rejects any notion that values can be read off the face of nature or the data of research. Diversity in some human traits, such as height, can be plotted on a bell-shaped curve. Those abnormally short might then be treated without being enhanced, if they are only brought up to the middle of the curve. However, for any trait, a technology that brings up those below average will in time have the effect of moving the whole of the curve, redefining the average and normal. If a statistical norm becomes the guide for intervention, it becomes a floating norm, unable to serve also as a moral norm.

If theology wishes to use the language of therapy and enhancement, it should not attempt to find a purely anthropological or natural basis for this distinction. Seen theologically, human beings are creatures whose meaning and destiny are only understood in relation to the Creator. While much can be learned about the human from scientific studies, the normative meaning of human nature is not found in biological and anthropological studies of the human in isolation but from an awareness of the human in relationship, first to other creatures but ultimately to God. From this perspective, enhancement is not the primary concern, at least not in Christian theology. The central theological question raised by technologies that modify humans is whether they serve the purposes of God in creating and renewing or redeeming the creation.

Viewed from this religious perspective, human beings are known to be caught up in a life-long process of improvement or enhancement or (to use a more traditional term) perfection. James F. Keenan, S. J., reviews some of the Christian literature on the theme of perfection. While it is true that Christians (most notably Protestants) have disagreed about the meaning and the process of attaining perfection, nearly all have agreed that moral and spiritual growth is central to the Christian experience. Keenan relates this theological history to the current debate over enhancement, noting that "the problem lies not with the question of *whether* we should pursue perfection, but rather *what perfection* we are pursuing."[16] In other words, there are various forms of perfection and enhancement, and the task of theological ethics is to distinguish between proper and improper forms.

For example, enhancement might be pursued for its own sake, simply to see how far a specific trait might be improved. Or it might be pursued for the sake of power and injustice, either so that the enhanced might dominate all others or so that they might be enhanced in their willingness to serve or to please others without hesitation. These things might be called enhancement, but it is their final purpose rather than their technological means that makes them morally objectionable. The same technology might be used to enhance other traits for other purposes, purposes that might even be regarded as noble, according to Keenan: "If we were willing to distinguish enhancement for itself and for oppressive power from enhancement for some more noble purpose, what would that latter purpose be? Here we are getting into the question of an anthropological vision for the human,"[17] something that is almost entirely missing from the contemporary world.

The challenge for theology, therefore, is not to search for a line between therapy and enhancement, but to seek to offer a vision of the future of humanity. In part, the secular preoccupation with the therapy–enhancement distinction may be seen as sign of growing uneasiness about a situation in which we have expanded powers to modify what we no longer think we understand and the future of which is open to unlimited interpretations. Technological transformation of humanity is possible precisely at a time when we no longer have a common view of human nature, and because we lack such a view we feel we have to permit everything we can imagine. No wonder we human beings, who live "in the absence of an adequate anthropological goal,"[18] are uncertain about how to assess the moral legitimacy of our enhancement projects. A task of increasing urgency is for theology to address this absence.

Perhaps with too much theological confidence, Keenan holds that "faith provides, then, an orientation to all reality and helps shape a theological anthropology that serves as the hermeneutical key with which to unlock the meaning of normative human nature. Thus faith helps guide our determination of human perfection."[19] Whatever truth there may be to this claim is diminished because Christian theologians do not agree on the shape of theological anthropology, largely because they disagree on how to relate the human and the divine. For too many

Christians even today, the starting point is an unrevised view of creation, which sees human beings as literal descendants of Adam and Eve and therefore as biologically inviolable because they are exactly the way God wants them to be. Also starting from creation, but with a full appreciation of the revisions to the doctrine that are necessary in light of biological and cultural evolution, Philip Hefner argues that human beings are best understood as God's co-creators, conscious subjects who act cooperatively with God in advancing the project of creation. Hefner suggests that co-creation be qualified somewhat to avoid the notion of human equality with God.[20] He suggests that we speak of ourselves as "created co-creators,"[21] creatures through whom God creates.

Some Christian theologians ground their understanding of the God–human relationship on the uniquely Christian doctrine of the incarnation. Most often this leads to a conclusion that is conservative in regard to technology. After all, if God takes up human form in one precise way—that of the humanity of Jesus Christ—then any departure from that form is loss rather than gain. Strictly speaking, according to this view there can be no such thing as genetic enhancement, because every genetic change is a move away from God's intent. According to Andy Crouch, the question for Christians is simple: Christ's humanity or technologically enhanced (and therefore diminished) humanity: "Do we want his life? Or do we want technology's alluring facsimile?"[22]

Yet another option is to ground the divine–human relationship in the future, focusing on what God is doing to renew or even to transform the creation, possibly by intending to create a successor form of humanity through our technology. According to Ted Peters, "a theology of continuing creation looks forward to the new . . . [and] is realistic about the dynamic nature of our situation. Everything changes."[23] From this follows an ethics, Peters says, that "denies that the status quo defines what is good, denies that the present situation has an automatic moral claim to perpetuity."[24] Instead, Peters insists, "the concept of the created co-creator we invoke here is a cautious but creative Christian concept that begins with a vision of openness to God's future and responsibility for the human future."[25]

This spectrum of Christian views may be compared with Judaism, which tends to view creation as standing in need of repair. Our duty as human beings is to share in the work of *tikkun olam*, or repairing creation. In chapter 2, Elliott Dorff explores the relationship between this traditional theme and recent technology. To the extent that germline modification makes new forms of healing or repair possible, it is our duty to advance the development of the technology that might make it possible to proceed safely. Of course it is possible to go too far with the mandate for healing and to do all sorts of evil in the name of good, claiming to do so with a divine blessing. The fear that human technology will go too far or that it will usurp the Creator's role is not dismissed, but neither is it held by Jewish scholars as strongly as it typically is by their Christian counterparts. In this context the comment of Laurie Zoloth is illuminating: "In Jewish theology, the case for the dangers of usurpation of this role is weak (not absent, but weak) and the case for active imitation of God's role is made strongly. Humans are mandated to use and control the natural world actively, to act as partners in God's creation, and to do *tikkun olam*, to repair the world."[26]

Applying these themes directly to the question of Judaism and human genetic enhancement, Jeffrey H. Burack writes: "Accepting covenantal responsibility does not argue against enhancement per se. It suggests, however, that we must always weigh seriously the value of undertaking an intervention against the profoundly unknowable potential ramifications in future generations. . . . We should take seriously an intervention that might make people more generous, for its potential to enhance *tikkun olam*."[27] This suggests, in much the same way as Keenan, that enhancement is not intrinsically objectionable. Our task is not to avoid enhancement, but to distinguish good and bad forms of enhancement.[28]

In the end, however, the distinction between appropriate and inappropriate uses of this technology or between good and bad forms of enhancement is best seen, not as a distinction within the technology, but within the technologist. Just as therapy cannot be distinguished sharply from enhancement, so it is impossible to label certain uses of enhancement as acceptable while others are considered unacceptable. In every case, the intention of the person using the technology comes clearly into play as the central feature. From the point of view of public policy, this is clearly a problem because the intention in the mind of someone using a technology cannot be discerned publicly or taken into account in a practical way. For religion, however, intention is central in many ways to the question of the rightness of an act. For this reason, one of the greatest contributions of religion to the public discussion of these technologies is to encourage reflection on our intentions, examination of our moral purposes, and moral preparation for the task of living with ever-increasing technological powers.

## Living with Technologies of Human Enhancement

Even if we wanted to, it is not clear we could still prevent the development of germline modification technologies. Nor is it likely we will ever find a clear and enforceable distinction between germline therapy and enhancement. From this, some will despair in the face of what they will consider a technological imperative, as if technology were a power unto itself, a power no longer under human control. In one respect this is true. Technology is not something we control, at least not in the sense of making grand decisions about its development. It is too late to turn it off.

Given enough time, the technologies of human germline modification will be developed. Technological hurdles will be overcome. More important, moral inhibitions and legal restrictions will not prevent the development and use of these technologies for therapeutic purposes and, in time, for enhancement. Some persons, of course, will try to outlaw this technology. They have succeeded already in some nations and they will probably succeed elsewhere, but it is unimaginable that they will succeed everywhere. For those who oppose the development of these technologies, the unpleasant truth is that failure to stop this development everywhere is failure to stop it at all. At most, isolated bans will only slow what cannot be stopped.

If the development of this technology is inevitable, it will also be incremental, based on many small steps rather than major advances. The future will come at us a bit at a time, like tenths of miles on a superhighway, too small to notice. Occasionally of course there are moments that mark major transitions. The announcement of the birth of Dolly, the cloned sheep, was such a moment. The general public, which rarely pays attention except for the occasional Dolly-like announcement, develops a distorted picture of the usual path of technological development. True breakthroughs are rare. Slow, painstaking work is more typical. Any technology as complex as human germline modification, which really must be considered as a suite of convergent technologies, advances by small steps across a broad front. This makes it all the more difficult to interpret or regulate it.

Furthermore, we should expect that the decisions we human beings will make about germline modification will not for the most part be grand or global political decisions, but isolated, small-scale decisions made in clinics or laboratories around the world, often by people who have almost no sense of the broader context of their choices. Incremental advances in technology will be met at every step by incremental moral decisions to accept or reject them. In some nations, decisions against germline modification are already in place, and perhaps a few more nations will put the question on a ballot or bring the matter to the legislature. What is not imaginable is a global plebiscite amounting to a decision that results in an enforceable ban. Far more likely are small, local, micro-decisions, thousands of them or more, which when aggregated and averaged will become our collective decision.

This is all the more reason, then, to focus our attention on ourselves rather than on our technology. Or more precisely, we ought to focus on ourselves as technologized beings, creators and users of technology who are to some extent our own creations. And so we must ask, what kind of people are we, who populate this transitional present moment and make decisions that will shape the lives of those to come? If we can no longer control the vast and rapid expanse of technology, we can at least aspire to control ourselves in the presence of technology. We can decide what kind of person each of us will be in a world of technology. It is therefore to ourselves that we should turn in order to probe our vulnerabilities to the new forms of temptation that technology presents, not in despair but in hope, not in desperation but to devote ourselves and our communities to the cultivation of new virtues and attitudes for a technological age, and not with delusions of control but with yearnings for renewed compassion and kindness.

The technologies of human modification pose new risks, not just of failure or unintended consequences, but also of distorting human relationships. Persons with disabilities, for example, are concerned that a wide range of genetic technologies will contribute to increased intolerance.[29] At the very least, these concerns should prompt a moral examination of why we use these technologies and in particular why we might turn to the future technologies of human germline modification. Is it possible to use these technologies without at the same time rejecting the very existence of those who live with conditions the technology is meant to change? Can we embrace a technology meant to avoid genetic conditions while at the same time fully welcoming those who may live with the very same conditions? If it is possible, it will require moral discernment, self-examination, cultural criticism, and concrete practice. Communities of faith in particular need to become more self-conscious in their practice of unconditional acceptance.

Another risk posed by human germline modification is that future parents will use it, not to have a child free of a genetic propensity to a disease, but to have a child with traits they find particularly attractive. This fear can be expressed in several ways. First, these future parents, thinking they can design the child they want, might think that they love this child for its traits and not for its unique and often perplexing existence. They will design the child they want rather than love the child they are given. They will come to see the child as a product of that design rather than a person full of surprises. When, in spite of all the engineering that might someday be developed, the child acts in unexpected ways, they might be even less prepared than ordinary parents to love the child unconditionally, through all the disappointments and demands of parenting. They might even think about the price they paid to design such a child. Human germline enhancement, if successful, will be expensive, and parents can be expected to think about whether they have gotten what they feel they paid for when they sought to enhance their future child's capabilities.

Over against all these possibilities, some fanciful, some all too likely, we human beings must hold on to what we value most in good parenting, even while recognizing how these technologies pose new threats to these values. Even parents who refuse to use these technologies cannot avoid being affected in their attitudes toward parenting by these emerging technical possibilities. Parents who cannot afford these technologies likewise will be affected as well, if only in envy of what they could not afford for their newly disadvantaged children.

Parents who take up these technologies and find themselves living with a genetically enhanced child—whatever that might mean—will feel most acutely the temptation to regard the child as something other than a child. It remains to be seen whether they will see the new life as a product of design rather than a person, or as an expensive purchase, or even as their genetic superior destined to exceed them in every respect. The value that must be defended, in the face of technologies that might erode it, is unconditional love and acceptance of a child as a gift, unexpected and free in personality and in a unique balance of traits. While technology might tempt us to think otherwise, it is unlikely that any technology will truly change a child into anything other than a human child.

Another risk posed by these technologies is that people will feel driven to embrace them out of the fear that unless they do so, their children will be put at a competitive disadvantage. The use of performance-enhancing drugs in competitive sports may in time be seen as nothing but a prelude to a much larger crisis. If it becomes possible to use human germline modification to enhance the cognitive abilities of our offspring, parents who fail to do so will be forced to recognize that their children will live, not so much with a disadvantage but without an advantage, perhaps as "unadvantaged." Whether these parents opt out on principle or for lack of financial resources, they will no doubt resent the circumstances that would appear to put their children on a track of mediocrity. Parents who use these technologies, regardless of their true motives, will know that their willingness to take up the use of these technologies is framed by a larger context of future competition.

It should be clearly pointed out here that competition per se is not the problem. The concern is over the way in which technology redefines and perhaps even eliminates true competition and the anxiety this induces in the individual competitor or in the parents of future competitors. Those concerned to keep drugs out of sports do so because they value competitive sports. Competition in real life, however, is far more fluid and uncontrolled, and for precisely these reasons it has become an engine of creativity, especially in recent decades. The question that requires more reflection is whether human germline enhancement, for example of cognitive abilities, will distort or enrich this competition. On the individual level, we might resent living in the presence of others who are enhanced beyond our own capacities. On the other hand, our individual lives are greatly enriched in every way, culturally and materially, precisely because we live among others with greater gifts.

All these risks point to the potential for these new technologies to tear at human social solidarity and to exacerbate inequities. The grand challenge before us today is to learn to live with these technologies, fully aware of the risks, actively engaged in countering their tendencies, and genuinely committed to unconditional social inclusion of all human lives, technologically modified or not. Those we modify will need more than our modifications. They will need our humanity and our face-to-face encounters. We will have questions about them, and they will have questions about themselves, perhaps wondering if they are so different after all.

We are entering an age not just of new technology but of technologized humanity, people whose identities and capacities will increasingly be shaped by multiple technologies. Technologized people will be feared as dangerous, envied as superhumans, shunned as anomalies, and followed as great leaders, probably all at once. What they will need most from us, and what we will need from them, is human engagement and support, with rituals to mark both our natural milestones and our technological modifications, new narratives of transformation to sustain continuity of character, and visions of a spiritualized technology that subsumes our heightened capacities in the purpose of higher meanings.

## Notes

1. LeRoy Walters and Julie Gage Palmer, *The Ethics of Human Gene Therapy* (New York: Oxford University Press, 1997), p. 82.

2. Jürgen Habermas, *The Future of Human Nature* (Cambridge, UK: Polity Press, 2003), p. 68.

3. Habermas, *Human Nature*, p. 69.

4. Habermas, *Human Nature*, p. 97.

5. Albert Moraczewski, "Germ-Line Intervention and the Moral Tradition of the Catholic Church," in Audrey R. Chapman and Mark S. Frankel, eds., *Designing our Descendants: The Promises and Perils of Genetic Modifications*

(Baltimore: Johns Hopkins University Press, 2003), pp. 199–211 at p. 210; italics in the original.

6. Edwin C. Hui [Xu Zhi-Wei], *At the Beginning of Life: Dilemmas in Theological Bioethics* (Downers Grove, IL: InterVarsity Press, 2002), p. 231.

7. Hui, *Beginning of Life,* p. 231.

8. Sondra Wheeler, "A Theological Appraisal of Parental Power," in Chapman and Frankel, eds., *Designing our Descendants*, pp. 238–251 at 250.

9. Wheeler, "A Theological Appraisal," pp. 250–251.

10. United Methodist Church, *Book of Discipline of the United Methodist Church* (Nashville, TN: United Methodist Publishing House, 1992), pp. 97–98.

11. See John Saward, *Redeemer in the Womb: Jesus Living in Mary* (San Francisco: Ignatius Press, 1993); and David Albert Jones, *The Soul of the Embryo: An Enquiry into the Status of the Human Embryo in the Christian Tradition* (London: Continuum, 2004), pp. 125–140.

12. Lee Silver, *Remaking Eden: How Genetic Engineering and Cloning will Transform the American Family* (New York: Avon Books, 1997).

13. Karl Rahner, "The Problem of Genetic Manipulation," in *Theological Investigations.* vol. 9, trans. G. Harrison (New York: Seabury, 1966), pp. 205–222, at 247.

14. Audrey Chapman, "The Implications of Inheritable Genetic Modifications for Justice," in Chapman and Frankel, eds., *Designing our Descendants*, pp. 131–155 at 152.

15. Brent Waters, *From Human to Posthuman: Christian Theology and Technology in a Postmodern World* (Hampshire, UK: Ashgate 2006), p. 37.

16. James F. Keenan, S. J., "'Whose perfection is it anyway?': A virtuous consideration of enhancement," *Christian Bioethics* 5.2 (1999), 104–120 at 105.

17. Keenan, "Whose perfection," p. 114.

18. Keenan, "Whose perfection," p. 114.

19. Keenan, "Whose perfection," p. 117.

20. For criticisms of the theme of co-creation, see Stanley Hauerwas, "Work as Co-Creation: A Critique of a Remarkably Bad Idea," in John W. Houck and Oliver F. Williams, eds., *Co-Creation and Capitalism: John Paul II's* Laborem Exercens (Washington, DC: University Press of America, 1983), pp. 42–58.

21. Philip Hefner, "The Evolution of the Created Co-Creator," in Ted Peters, ed., *Cosmos as Creation: Theology and Science in Consonance* (Nashville, TN: Abingdon Press, 1989), pp. 211–234.

22. Andy Crouch, "When backward is forward: Christmas may be the best argument against genetic enhancement," *Christianity Today* (December 2004), p. 66.

23. Ted Peters, *Playing God? Genetic Determinism and Human Freedom* (New York: Routledge, 1997), p. 155.

24. Peters, *Playing God?*, p. 155.

25. Peters, *Playing God?,* p. 156.

26. Laurie Zoloth, "Uncountable as the Stars: Inheritable Genetic Intervention and the Human Future—A Jewish Perspective," in Chapman and Frankel, eds., *Designing our Descendants*, pp. 212–237 at 218.

27. Jeffrey H. Burack, "Jewish reflections on genetic enhancement," *Journal of the Society of Christian Ethics* 26:1 (2006): 137–161 at 157.

28. The suggestion that there are moral distinctions to be made among various forms of enhancement is explored briefly in Celia Deane-Drummond, "God, the Transhuman Future and the Quest for Immortality," in Celia Deane-Drummond and Peter Manley Scott, eds., *Future Perfect? God, Medicine and Human Identity* (London: T. & T. Clark, 2006), pp. 168–182 at 180.

29. See John Swinton and Brian Brock, eds., *Theology, Disability and the New Genetics: Why Science Needs the Church* (London: Continuum, 2007).

# Suggestions for Further Reading

Bryant, J. and J. Searle (2004). *Life in Our Hands: A Christian Perspective on Genetics and Cloning*. Leicester, UK: Intervarsity Press.

Cahill, L. S. (2004). *Bioethics and the Common Good*. Milwaukee, WI: Marquette University Press.

Cahill, L. S. (2005). *Genetics, Theology, and Ethics: An Interdisciplinary Conversation*. New York: Crossroad Publishing.

Cahill, L. S. (2005). *Theological Bioethics: Participation, Justice, and Change*. Washington, DC: Georgetown University Press.

Cameron, N. M. D. S. (1992). *The New Medicine: Life and Death after Hippocrates*. Wheaton, IL: Crossway Books.

Chapman, A. R. (1999). *Unprecedented Choices: Religious Ethics at the Frontiers of Genetic Science*. Minneapolis: Fortress Press.

Chapman, A. R. and M. S. Frankel (2003). *Designing Our Descendants: The Promises and Perils of Genetic Modifications*. Baltimore: Johns Hopkins University Press.

Cole-Turner, R. (1993). *The New Genesis: Theology and the Genetic Revolution*. Louisville: Westminster John Knox Press.

Cole-Turner, R. (2001). *Beyond Cloning: Religion and the Remaking of Humanity*. Harrisburg, PA: Trinity Press International.

Deane-Drummond, C. (2003). *Brave New World?: Theology, Ethics and the Human Genome*. London: T & T Clark.

Deane-Drummond, C. (2006). *Genetics and Christian Ethics*. Cambridge, UK: Cambridge University Press.

Deane-Drummond, C. and P. Scott (2006). *Future Perfect?: God, Medicine and Human Identity*. London: T & T Clark International.

Dorff, E. N. (2005). *The Way into Tikkun Olam (Repairing the World)*. Woodtsock, VT: Jewish Lights Publishing.

Dorff, E. N. and L. E. Newman (1995). *Contemporary Jewish Ethics and Morality: A Reader*. New York: Oxford University Press.

Engelhardt, H. T. (2000). *The Foundations of Christian Bioethics*. Lisse [the Netherlands]; Exton, PA, Swets & Zeitlinger Publishers.

Engelhardt, H. T. (2006). *Global Bioethics: The Collapse of Consensus*. Salem, MA: M & M Scrivener Press.

Evans, J. H. (2002). *Playing God?: Human Genetic Engineering and the Rationalization of Public Bioethical Debate*. Chicago: University of Chicago Press.

Fukuyama, F. (2003). *Our Posthuman Future: Consequences of the Biotechnology Revolution*. New York: Picador.

Habermas, J. (2003). *The Future of Human Nature*. Cambridge, UK: Polity.

Heap, B. (2004). *Pastoral Implications of the New Genetics*. Carlisle, Cumbria: Partnership/Paternoster Press.

Hefner, P. J. (1993). *The Human Factor: Evolution, Culture, and Religion*. Minneapolis: Fortress Press.

Hefner, P. J. (2003). *Technology and Human Becoming*. Minneapolis: Fortress Press.

Hui, E. C. (2002). *At the Beginning of Life: Dilemmas in Theological Bioethics*. Downers Grove, IL: InterVarsity Press.

Jonas, H. (1984). *The Imperative of Responsibility: In Search of an Ethics for the Technological Age*. Chicago: University of Chicago Press.

Junker-Kenny, M. and L. S. Cahill (1998). *The Ethics of Genetic Engineering*. London: SCM Press.

Mackler, A. L. (2003). *Introduction to Jewish and Catholic Bioethics: A Comparative Analysis*. Washington, DC: Georgetown University Press.

McKibben, B. (2003). *Enough: Genetic Engineering and the End of Human Nature*. London: Bloomsbury.

Mitchell, C. B. (2006). *Biotechnology and the Human Good*. Washington, DC: Georgetown University Press.

Parens, E. (1998). *Enhancing Human Traits: Ethical and Social Implications*. Washington, DC: Georgetown University Press.

Peters, T. (2003). *Playing God?: Genetic Determinism and Human Freedom*. New York: Routledge.

Peterson, J. C. (2001). *Genetic Turning Points: The Ethics of Human Genetic Intervention*. Grand Rapids, MI: W.B. Eerdmans Publishing.

Ramsey, P. (1970). *Fabricated Man; the Ethics of Genetic Control*. New Haven: Yale University Press.

Reichenbach, B. R. and V. E. Anderson (1995). *On Behalf of God: A Christian Ethic For Biology*. Grand Rapids, MI: Eerdmans with the Institute for Advanced Christian Studies.

Shannon, T. A. (2000). *Made in Whose Image?: Genetic Engineering and Christian Ethics*. Amherst, NY: Humanity Books.

Shannon, T. A. and J. J. Walter (2003). *The New Genetic Medicine: Theological and Ethical Reflections*. Lanham, MD: Rowman & Littlefield Publishers.

Shinn, R. L. (1996). *The New Genetics: Challenges for Science, Faith, and Politics*. Wakefield, RI: Moyer Bell.

Silver, L. M. (1997). *Remaking Eden: Cloning and Beyond in a Brave New World*. New York, Avon Books.

Song, R. (2002). *Human Genetics: Fabricating the Future*. London: Darton Longman and Todd.

Stock, G. (2003). *Redesigning Humans: Choosing Our Genes, Changing Our Future*. Boston: Houghton Mifflin.

Stock, G. and J. H. Campbell (2000). *Engineering the Human Germline: An Exploration of the Science and Ethics of Altering the Genes We Pass to Our Children*. New York: Oxford University Press.

Walter, J. J. and T. A. Shannon (2005). *Contemporary Issues in Bioethics: A Catholic Perspective*. Lanham, MD: Rowman & Littlefield Publishers.

Waters, B. (2001). *Reproductive Technology: Towards a Theology of Procreative Stewardship*. London: Darton, Longman and Todd.

Waters, B. (2006). *From Human to Posthuman: Christian Theology and Technology in a Postmodern World*. Aldershot, UK: Ashgate Publishing.

Waters, B. and R. Cole-Turner (2003). *God and the Embryo: Religious Voices on Stem Cells and Cloning*. Washington, DC: Georgetown University Press.

# Contributors

**Lisa Sowle Cahill** is the J. Donald Monan Professor of Theology at Boston College in Massachusetts. Her most recent books are *Theological Bioethics: Participation, Justice and Change* (winner of the 2006 first place award for outstanding theology books from the Catholic Press Association) and *Bioethics and the Common Good.*

**Nigel M. de S. Cameron** is the director of the Center on Nanotechnology and Society at the Illinois Institute of Technology in Chicago and president of the Institute on Biotechnology and the Human Future. His chief interest lies in the implications of emerging technologies for policy and human values. He is the co-editor of *Nanoscale: Issues and Perspectives for the Nano Century*, and he is currently working on his next book, *Choosing Tomorrow.*

**Ronald Cole-Turner** holds the H. Parker Sharp Chair in Theology and Ethics at Pittsburgh Theological Seminary in Pennsylvania. He is the author of *The New Genesis: Theology and the Genetic Revolution,* co-author of *Pastoral Genetics: Theology and Care at the Beginning of Life*, and editor of various collections on emerging biotechnologies.

**Celia Deane-Drummond** holds a chair in theology and the biological sciences at the University of Chester in the United Kingdom and is director of the Centre for Religion and the Biosciences, which she founded in 2002. Her most recent publications include *Wonder and Wisdom: Conversations in Science, Spirituality and Theology*, *Genetics and Christian Ethics*, and *Future Perfect: God, Medicine and Human Identity*, edited with Peter Scott.

**Amy Michelle DeBaets** is a Ph.D. student in religious ethics at Emory University, Atlanta, Georgia. She received her M.Div. and Th.M. degrees from Princeton Theological Seminary and has taught bioethics at Cornell University, Ithaca, New York.

**Elliot Dorff** is rector at the University of Judaism, Los Angeles, California, where he also holds the Sol and Anne Dorff Distinguished Service Professorship in Philosophy and serves as co-chair of the bioethics department. Recent publications include *Love Your Neighbor and Yourself: A Jewish Approach to Modern*

*Personal Ethics* (2003), *The Unfolding Tradition: Jewish Law After Sinai* (2005), and *The Way Into Tikkun Olam (Fixing the World) (2005).*

**H. Tristram Engelhardt, Jr.** is a professor with Department of Philosophy, Rice University, Houston, Texas and professor emeritus, Department of Medicine, Baylor College of Medicine, Waco, Texas. He is the editor of the *Journal of Medicine and Philosophy* and the senior editor of the journal *Christian Bioethics.* He is also the editor of the Philosophy and Medicine book series and the Philosophical Studies in Contemporary Culture book series. His recent volumes and *Global Bioethics: The Collapse of Consensus* and *Foundations of Christian Bioethics.*

**Thomas A. Shannon** is professor emeritus of religion and social ethics in the Department of Humanities and Arts at Worcester Polytechnic Institute, Worcester, Massachusetts. While there, he was the Paris Fletcher Distinguished Professor of the Humanities from 1994 to 1996. Currently he holds the Paul McKeever Chair of Moral Theology at St. John's University, New York. He is the author or co-author or editor of more than thirty books and forty articles in bioethics and Roman Catholic social justice.

**James J. Walter** is the Austin and Ann O'Malley Professor of Bioethics and chair of the Bioethics Institute at Loyola Marymount University in Los Angeles. His most recent publications include *The New Genetic Medicine: Theological and Ethical Reflections*; *Contemporary Issues in Bioethics: A Catholic Perspective;* and *Artificial Nutrition and Hydration and the Permanently Unconscious Patient: The Catholic Debate.*

# Index

Access to technology, 157–162
Agar, Nicholas, 139–13
Agius, Emmanuel, 91–20
Albert the Great, Saint, 182
American Association for the Advancement of Science (AAAS), 21, 27–42
Anderson, V. Elving, 226, 140–24
Anderson, William French, 131, 138–4, 138–10, 139–13, 142–48, 142–49
Angell, Marcia, 159–160, 162, 165–41, 165–42, 165–43, 165–44, 165–45, 165–46, 165–47, 166–53
Aquinas. *See* Thomas Aquinas, Saint
Aristotle, 187
Athanasius, Saint, 90–8, 91–19
Augustine of Hippo, Saint, 180–181, 194–195
Autonomy, 212

Babel, the Tower (of the biblical book of Genesis), 38–39, 106–107
Bainbridge, William Sims, 116–3
Barbour, Ian G., 139–16, 140–30, 141–34, 141–44, 142–54, 142–56
Barnet, Robert J., 165–37
Barritt, J. A., et al., 25–5, 26–36
Bartholomew, Ecumenical Patriarch of Constantinople, 81, 90–11
Baruch, Susanna, et al., 2, 4, 17, 24–2, 25–4, 26–34, 116–4
Basil the Great, Saint, 83–85, 91–15, 91–16, 91–17
Bayertz, Kurt, 163–2
Begley, Sharon, 116–1
Behavioral Research, 139–13, 139–14, 140–25
Bernstein, Ellen, 50–29
Bioethics, 4–6, 212–213
  arguments for germline modification, 101–103, 122, 146 187, 202
  arguments in opposition to germline modification, 121–122, 146, 156–157, 178, 187
Blaese, R. Michael, 27–38, 27–39
Bonaventure, Saint, 182
Brock, Brian, 223–29
Bryant, J., 200–62, 225
Buchanan, Allen et al., 25–7, 164–21, 166–52
Bunyan, John, 100
Burack, Jeffrey H., 216, 223–27
Busuttil, Salvino, 91–20

Cahill, Lisa Sowle, 24, 162, 209, 225–226, 140–27, 166–55
Cain and Abel (of the biblical book of Genesis), 106, 108
Callahan, Daniel, 141–42
Cameron, Nigel M. de S., 15, 23, 210, 225, 117–5
Campbell, J. H., 227
Carter Center, 166–54

Catholic Health Association of the United States, 139–13, 166–54
Catholic Bishops, United States Conference, 139–16
Catholic Magisterium or official teaching, 30, 52–55, 208
Catechism of the Catholic Church, 52, 55–57, 69–2, 69–3, 69–5, 69–10, 69–14, 69–15, 69–18, 69–19, 70–28
Congregation of the Doctrine of the Faith, 14, 52, 54–56, 69, 122–123, 26–33, 69–1, 69–7, 69–12, 69–13, 69–16, 69–17, 70–27, 70–30 70–46
International Theological Commission, 14, 64, 124, 26–31, 70–39, 139–22
Pontifical Academy for Life, 52–53, 57–58, 69–4, 69–6, 69–20, 70–22, 70–23, 70–24, 70–25, 70–26, 70–29, 70–34, 70–35, 70–36, 70–42
Pope Benedict XVI, 26–31
Pope John Paul II, 14, 54, 57, 60, 66, 123, 26–31, 26–32, 69–8, 69–21, 70–43, 139–13, 139–19, 139–20, 139–21, 141–36, 142–51
Holy See, 26–33, 69–11, 70–31, 70–33
Vatican II, 77, 89–3
Catholicism, 13–14, 22, 24–25, 52–69, 203, 206
and cloning, 61–62
and the common good, 124
and embryonic stem cell research, 62–63, 206–209
and germline modification, 57–58, 63–65, 124, 206–209
and human subject research, 53–57, 60
medical ethics, 51, 53–56, 123, 47–3
and natural law, 123
theology, 13
and xenotransplantation, 58–60
Center for Genetics and Society, 166–54
Cessario, Romanus, 198–37, 199–53
Chadwick, Derek, et al., 142–53
Chan, A. W. S., et al., 25–3
Chapman, Audrey, 210, 225, 27–42, 222–14
Charlesworth, Max, 142–53
Christian beliefs
and biotechnology, 1, 84–87, 132–133, 136–137
creation, 78–80, 96–98, 104, 125–129, 146–147, 150–151
eschatology, 97, 115–116, 134–136
fall into sin, 79, 86, 96–97, 129–131, 133–134, 168–172, 178, 213–217
and germline modification, 84–87, 107–108, 136–137, 205–217
grace, 174, 177
Holy Spirit, 77–78, 83, 174, 177, 183
human nature, 79–80, 85, 98–99, 103–105, 109, 127–129, 146–157, 167, 172
humans as co-creators, 98, 109–110, 116, 125, 128–129, 132, 151–153, 192, 215
incarnation of Jesus Christ, 79–80, 85, 96–97, 104, 113, 116, 125, 131–132, 207, 215
and medicine, 83–84, 86, 96–99, 132, 134
redemption, 132–134
responsibility for children, 85, 113–114, 219–221
responsibility for creation, 98–102, 114, 123, 125, 127–129, 135–136, 150–153
and social justice, 115, 145–146, 149–150, 153, 155–162, 209–211, 219–221
and technology, 83–84, 98–100, 106–108, 113–116, 132–134, 151–152

Chromosomes, artificial, 18
Clague, Julie, 153, 164–18, 197–25
Clarke, W. Norris, S.J., 141–44
Cloning, 61–62, 193, 195–196, 217
Cole, A. P., et al., 139–18
Competition, 220–221
Conscience, 168, 172, 178–183, 189, 194–196
 and errors, 181–182, 186–187
 as *synderesis*, 179–180, 185
 and virtue, 182
Cole-Turner, Ronald, 134, 156, 225, 227, 7–7, 26–24, 117–6, 138–12, 141–35, 142–60, 164–29, 196–4, 199–50, 199–58
Convention for the Protection of Human Rights and Dignity, Council of Europe, 87, 92–21
Coutts, Mary Carrington, 138–11
Cox, Harvey, 135, 142–61, 142–63, 143–63
Cox, Paul M., 142–58
Crick, Francis, 51
Crouch, Andy, 215, 222–22
Culver, Kenneth, 27–39, 200–71
Curran, Charles E., 140–28

Daniels, Norman, 155, 164–21, 164–22, 164–23, 164–24, 164–25, 164–26, 164–26
Davies, Brian, 180–181, 198–31
Davies, Michael, 89–3
de Chardin, Pierre Teilhard, 133, 142–57
de Wachter, Maurice A. M., 138–11
Deane-Drummond, Celia, 24, 225, 198–36, 198–38, 198–45, 200–69, 223–28
DeBaets, Amy Michelle, 15, 23, 210
"Designer babies," 1–2, 16, 202
Disabilities, 219
Doerflinger, Richard, 70–43
Dolly, the cloned sheep, 217
*Donum vitae*. *See* Catholic, Congregation for the Doctrine of the Faith
Dorff, Elliot N., 22, 216, 225, 47–1, 47–2, 47–4, 47–5
Drane, James, 47–3
Dunston, G. R., 131, 142–50

Eggs. *See* Oocytes
Eisenstein, J. D., 49–22
Ellul, Jacques, 132, 142–55
Elmer-Dewitt, Philip, 140–33
Embryo, human 17, 201–203
 dignity of, 66–68, 85
 individuality, 66–67
 and twinning, 66
Embryo Research, 53–57, 60–62, 65–68, 85, 123–124, 193–194
Engelhardt, H. Tristam, Jr., 22–23, 226, 89–3, 92–22, 92–23
Enhancement, see Therapy vs. enhancement
Ethical, Legal and Social Implications (ELSI). *See* Human Genome Project
Eugenics, 100, 114, 191–192
Evagrious the Solitary, 89–4
Evans, John H., 226, 26–24

Fletcher, Joseph F., 142–46
Frankel, Mark, 161–162, 27–42, 165–50, 166–51
Friedman, Theodore, 25–6, 139–13
Fuchs, Josef, S.J., 140–32
Fukuyama, Francis, 5–8, 82–83, 178, 226, 25–9, 25–10, 25–11, 25–12, 25–13, 25–17, 25–18, 91–12, 91–13, 196–2

GATTACA, 2, 192
Gelsinger, Jesse, 102
Gene repair, 19, 27–38, 27–39
Gene targeting (or homologous recombination), 20, 27–40, 27–41
Gene therapy. *See* Somatic cell gene therapy
Genetics and Public Policy Center, 2, 4, 17, 24–2, 25–4, 26–34, 116–4

Germline modification
definition, 4, 16–17, 20–21, 145, 196–1
inadvertent or unintentional, 20–22, 27–42, 27–43
safety, 206
techniques, 2–4, 17–20, 192–195
Global Fund to Fight AIDS, 159, 165–38
Gregory Palamas, Saint, 89–4
Gustafson, James M., 140–30, 141–39

Habermas, Jürgen, 6–7, 9, 82–83, 203, 226, 25–19, 25–20, 25–23, 91–14, 221–3, 221–4
Hall, Douglas John, 152, 163–12, 163–13
Häring, Bernard, 140–31, 197–27
Hauerwas, Stanley, 222–20
Hayes, Zachary, 140–30
Heap, Brian (Sir), 226
Hefner, Philip, 109, 215, 226, 117–15, 117–16, 117–17, 141–40, 222–21
Hegel, Georg W. F., 90–6
Heim, Rudolf, 198–39
Heschel, Abraham Joshua, 48–15
Heyd, David, 163–3, 164–19
Hilkert, Mary Catherine, O.P. 163–8
Hodge, Charles, 105, 117–12
Hogan, Linda, 180–182, 197–27, 197–29, 198–32
Hui, Edwin C. [Xu Zhi-Wei], 204, 226, 222–6, 222–7
Human as artifact, 107–108
Human dignity, 80, 94–95, 171
Human enhancement, 85–86, 94–96, 105, 110–113, 154–157, 210–221
Human freedom, 101–102, 109–110, 167–180, 183, 194–195
and knowledge, 168–171, 174, 179
transcendental freedom, 174–177, 179, 183, 191, 194–196
Human Genome Project, 51, 93, 119–120, 124, 189
Human germline genetic modification. *See* Germline modification
Human identity, 167, 177
and germline modification, 177
Human nature, 7, 87–88, 96–99, 105, 110–112, 145–157, 212–213
embodied, 148–150, 167
and genetics, 146–147, 189–190
Human power over nature, 100–102, 109–111
Human subjects research, 52–57, 60
Huxley, Aldous, 105, 111, 161

Image of God. *See* Christian beliefs, human nature
In vitro fertilization (IVF), 9, 15, 191
Inheritable genetic modifications. *See* Germline modification

Jacob (of the biblical book of Genesis), 40–41
Jakobovits, Immanuel, 49–21
Jans, Jan, 140–32
John Chrysostom, 91–9
Johnstone, Brian V., C.SS.R., 138–8
Jonas, Hans, 1, 24, 226, 24–1
Jones, D. Gareth, 141–38
Jones, David Albert, 222–11
Jones, Kenneth C., 89–3
Joy, Bill, 108, 117–14
Judaism, 29–46, 207, 215–216. *See also* Rabbis
and biotechnology, 1
Conservative, 31–32
doctrines, 32–46, 47–5
and germline modification, 39–40, 44–46, 216
and medicine, 34–36
method, 30–31, 47–4
Mishah, 36, 40, 42–43
Orthodox, 31, 38
Reform, 32
Talmud, 32–36, 39–44
*tikkun olam*, 36–38, 45, 215–216
Torah, 32, 35, 37–38, 41–44, 90–7

Juengst, Eric, 4, 121–122, 25–5, 26–37, 27–43, 139–17, 140–30, 143–65, 165–31
Junker-Kenny, M., 226
Justice, 16, 24, 115, 149–150, 155–162, 168, 184, 201, 209–211, 219–221

Kant, Immanuel, 89–2
Kass, Leon, 5–6, 25–14, 25–15, 25–16, 25–22
Keeler, Cardinal William H., 70–32
Keenan, James F., S.J., 153, 213–214, 216, 140–26, 163–17, 222–16, 222–17, 222–18, 222–19
Kolata, Gina, 116–2
Krimsky, Sheldon, 161, 165–48, 165–49
Kubota, H., et al., 27–41

Lammers, Ann, 141–41
Lappé, Marc, 138–9
Lee, Thomas F., 138–3, 138–6
Lewis, C. S., 99–101, 117–8, 117–9, 117–10
Little, Margaret Olivia, 165–32
Locke, John, 29
Love, 174

MacIntyre, Alasdair, 88, 92–24
Mackler, Aaron, 226, 47–3
Magisterium. *See* Catholic Magisterium
Mann, Thomas, 152, 163–15
Mantzarides, Georgios I., 91–19
Marcy, Marin E., 140–29
Market forces, 157–162
Mauron, Alen, 139–14, 139–14
McClaren, Anne, 196–3
McGleenan, Tony, 141–45
McKibben, Bill, 226, 118–25, 200–66
Medicare, 160
Messer, Neil, 199–5
Metz, Johannes B., 142–61
Mitchell, C. Ben, 226
Montesquieu, Claude, 29
Moraczewski, Albert, 203–204, 221–5
Murray, Thomas H., 165–40

National Council of the Churches of Christ, USA, 12, 98, 26–27, 117–6
National Nanotechnology Initiative of the National Science Foundation, 94, 116–3
Natural law, 179–180, 183
  Orthodox rejection, 79–80
Nelson, J. Robert, 138–12, 140–30
Nevins, Michael, 49–21
Newman, Lewis E., 225, 47–4
Noah and the Ark (of the biblical book of Genesis), 106–107
Nolan, Hugh J., 139–16
Nominalism, 172–174, 186
Nowak, R., 200–70

Oocytes, 17
Ooplasm transfer, 4, 19, 25–5, 26–36
Orthodox Christianity, 15
  definition, 73–79
  differences with western Christianity, 78
  doctrines, 79–82
  experience of God, 80–82
  and germline modification, 84–87
  and medicine, 83–86
  and philosophy, 82–83, 87–88
  role of creeds, 77–78
  role of tradition, 75–78

Page, Ruth, 163–16
Palmer, Julie Gage, 101–102, 202–203, 25–8, 117–11, 221–1
Parens, Erik, 4, 226, 25–5, 26–37, 27–43, 163–1, 164–30, 165–33, 165–34
Peters, Ted, 178, 215, 226, 141–40, 141–41, 197–24, 199–50, 223–23, 223–24, 223–25

Peterson, James C., 226
Phan, Kinh Luan, et al., 138–1
Philosophy and public debate, 5
  objections to germline modification, 5
  and religion, 5–7, 29, 82–83, 87–88, 212–213
Pinckaers, Servais, 172–173, 196–10, 196–11, 196–12, 196–13
"Playing God," 9, 23, 121, 124–125, 187, 195
Preimplantation Genetic Diagnosis (PGD), 202–203, 208
Protestantism, 11–13, 30, 204–205, 208–209
Prudence, 182–185, 189, 195
Public debate and religion, 1–2, 4–6, 217

Rabbis
  Akiba ben Joseph, 44
  Eliezer, 43, 90–7
  Hanina, 43, 90–7
  Hillel, 111, 47–9
  Karo, Joseph, 34–35, 49–24
  Maimonides, 45–46, 47–9
  Moses ben Nahman, 35
  Oshaya, 43
Rae, Scott B., 142–58
Rahner, Karl, 171, 174–177, 179, 183, 188–189, 191–192, 194–195, 209–210, 196–9, 197–15, 197–16, 197–17, 197–18, 197–19, 197–20, 197–21, 197–22, 197–23, 197–26, 199–51, 199–52, 199–54, 199–55, 199–56, 200–63, 200–64, 200–65, 222–13
Ramsey, Paul, 9–10, 16, 124, 132, 136, 151, 187–188, 195, 226, 25–21, 26–25, 139–23, 142–52, 143–64, 163–9, 198–44, 199–46, 199–47, 199–48, 199–49, 199–50
Reich, Warren T., 139–13
Reichenbach, Bruce R., 226, 140–24
Religion
  definition, 29
  opposition to germline modification, 10
  and public debate, 1–2, 5–8
  support for germline modification, 8–16
  value of religion, 8
Rifkin, Jeremy, 10, 25–28–24
Risk. *See* Safety
Roco, Mihail C., 116–3
Roth, Cecil, 49–21
Roth, Sol, 50–33

Safety, 59–60
Saward, John, 222–11
Scholasticism, 172
Scott, Peter, 225
Scully, Jackie Leach, 118–24
Searle, J., 200–62, 225
Sermon on the Mount, 174
Shannon, Thomas J., 13, 22, 151, 203, 206, 208, 226–227, 70–44, 70–45, 141–37, 141–43, 163–10, 163–11
Shinn, Roger Lincoln, 227
Siep, Ludwig, 148, 163–4, 163–5
Silouan the Athonite, Saint, 90–5
Silver, Lee, 115, 209, 227, 222–12
Smith, Cynthia, et al., 165–36
Somatic cell gene therapy, 17, 93–95, 119–120, 145
Song, Robert, 149, 227, 163–6, 163–7
Southern Baptist Convention, 13, 26–30
Sperm, 17
Sperm precursor cells, 17
Spielberg, Steven, 117–13
Stock, Gregory, 227
Supreme Court of the United States, 30, 48–1
Suzuki, Nobutaka, et al., 26–35
Swinton, John, 223–29

Symeon the New Theologian, Saint, 89–4
Szasz, Thomas, 157, 165–35

Temptation, 169–170
Therapy vs. enhancement, 15–16, 23, 85–86, 94, 95–96, 110–113, 120–121, 134, 154–155, 190, 195–196, 201–202, 204–205, 211–217
Thévoz, Jean-Marie, 139–14, 139–15
Thomas Aquinas, Saint, 172, 174, 179–187, 194–195, 196–14, 197–26, 197–28, 198–33, 198–34, 198–43
Thomism, 172, 174
Thomson, J. A., 27–40

United Methodist Church, 12–13, 206, 26–28, 26–29, 222–10
United Nations Educational, Scientific and Cultural Organization (UNESCO), 94–95
United Nations, 94–95, 159, 165–39
Universal Declaration of Human Rights, 54–55
U.S. President's Commission for the Study of Ethical Problems in Medicine and Biomedical and Behavioral Research, 139–13, 139–15, 140–25
U.S. President's Council on Bioethics, 110–112, 117–18, 117–19, 117–20, 117–21, 138–5, 199–57

Vardy, Peter, 185–186, 198–41
Vaux, Kenneth, 140–29
Vectors, viral and nonviral, 18
Verhey, Allen, 140–27
Vincent of Lerins, Saint, 89–5
Virtue ethics, 24, 168, 172, 177–179, 182–185
  and germline modification, 184–187, 189, 192
Vlachos, Hierotheos, 89–4
von Rad, Gerhard, 169

Walter, James J., 13, 23–24, 151, 202, 227, 140–28, 163–10, 163–11
Walters, LeRoy, 101–102, 202–203, 25–8, 117–11, 221–1
Waskow, Arthur, 50–29
Waters, Brent, 212, 227, 196–4, 222–15
Wathen, James F. 89–3
Watson, James D., 51, 119
Westberg, Daniel, 185, 198–35, 198–40
Westerman, Claus, 152, 169, 163–14, 196–5, 196–6, 196–7, 196–8
Wheeler, Sondra Ely, 104, 204–205, 117–23, 222–8, 222–9
William of Ockham, 172, 186
Williams, Esther, 89–4
Wilmut, Ian, 200–68
Wisdom, 183–185, 195–196
Wolter, Allan B., OFM, 70–44
World Council of Churches, 11–12, 26–26
World Health Organization, 155, 164–20

Xu Zhi-Wei. *See* Hui, Edwin C.

Zimmerman, Burke K., 138–7
Zoloth, Laurie, 44–45, 216, 50–45, 223–26
Zwaka, T. P., 27–40